An Introduction
To
Engineering
Measurements

An Introduction To Engineering Measurements

A. RICHARD GRAHAM

Department of Mechanical Engineering
Wichita State University

PRENTICE-HALL, INC., Englewood Cliffs, New Jersey

Library of Congress Cataloging in Publication Data

Graham, Archie Richard.
    An introduction to engineering measurements.

    Includes bibliographical references.
    1. Engineering instruments. I. Title.
TA165.G68   620'.004'4   74–5194
ISBN 0–13–482406–7

© 1975 by Prentice-Hall, Inc.
Englewood Cliffs, New Jersey

10 9 8 7 6 5 4 3 2 1

Printed in the United States of America

PRENTICE-HALL INTERNATIONAL, INC., *London*
PRENTICE-HALL OF AUSTRALIA, PTY, LTD., *Sydney*
PRENTICE-HALL OF CANADA, LTD., *Toronto*
PRENTICE-HALL OF INDIA PRIVATE LIMITED, *New Delhi*
PRENTICE-HALL OF JAPAN, INC., *Tokyo*

Contents

Preface

This book was written for a specific purpose: to give you, the reader, the ability to examine the specifications of a particular instrument and predict the way that the instrument will respond to various inputs. All the material in the text, with the exception of the last chapter, has been selected to accomplish this goal. In order to keep the book short and directed to the stated purpose, I have omitted some information that is ordinarily included in texts of this nature. Much of this is mentioned, with suitable reference, in the last chapter.

Why is it more important to be able to interpret specifications in terms of performance than to know, for instance, how a thermocouple or a bourdon–tube pressure gage or an oscilloscope operates? You may never see a thermocouple, or a bourdon tube pressure gage, or an oscilloscope. It is almost certain, however, that you will deal with data that result from the use of instrumentation.

How should one view this kind of data? It seems to me that the reaction is either complete distrust or absolute acceptance of the data as completely correct. Certainly neither is appropriate. I believe that on completing the study of the book you will be aware that any experimental results are subject to errors of various kinds. You should be able to anticipate these errors and to treat them appropriately.

The Contents indicates the organization of the book. First is a discussion of the various functions components of an instrument may perform. The following chapters explain the response to steady inputs and to inputs that are functions of time. Then follows an explanation of the various types of input signals that may be encountered. This material is then used in interpreting manufacturers' specifications. The book concludes with a chapter indicating directions that your future studies might take.

I have included a number of aids that will assist you in learning this material. Each chapter opens with a statement defining the purpose of the chapter. If you can perform the activity cited in this statement, you may be certain that you have satisfactorily mastered the material. You will find problems and questions in appropriate places to test yourself. As you read, you will also encounter questions. Don't skip them; answer them as well as you can. These are intended to draw your attention to particular points and to invite your active participation in the process of developing the subject. I have provided answers to each of these questions in the text so that you can compare answers with me.

Wichita, Kansas   A. Richard Graham

An Introduction To Engineering Measurements

Introduction

What is an instrument? Why should I be interested in instrumentation? What should I know about instruments? Perhaps these questions have occurred to you and are concerning you as you begin your study of this book. Let us consider these questions and a few others in the following paragraphs.

First, just what is an instrument? We will consider an *instrument* to be a device which is able to sense some physical parameter (pressure, temperature, velocity, etc.) and convert it to a form which is quantitized and can be interpreted by a human observer. Instruments may be quite simple, such as the liquid-in-glass thermometer or extremely complex, such as the devices to sense the physiological reactions of man during space flight. We frequently call instruments that are complex and composed of a number of components *instrumentation systems*.

Now, why should you be interested in instrumentation? Certainly most engineers do not become specialists in this field. In fact, many never have direct contact with instruments (other than those all of us encounter daily). However, many of us are required at various times to design experiments or specify performance tests which require certain measurements. We also find it necessary to interpret the output from instruments. That is, we have data which supposedly represent some parameter of a system. Just how reliable is this information? Does it require correction? If so, what correction is necessary and how can it be accomplished? In other words, we must be able to determine how closely the instrument output reflects the value of the variable that is being measured. We are interested in determining how well the instrument is performing.

What can we learn about the performance of a particular instrument and what is the source of this information? The manufacturer of the instrument furnishes this information in the form of specifications. These specifications give values of selected characteristics of the instrument that, in turn, describe the way the instrument responds to certain types of input. What are these characteristics? How can you be sure the instrument will respond adequately to the inputs you expect to encounter? What is the proper interpretation of the specifications of an instrument?

By the time you complete this book you should be able to accomplish the following:

> *Given a manufacturer's specifications, you will be able to describe how an instrument will respond to a static input and to various time-dependent inputs.*

1

Functional Elements and Operating Characteristics

We are primarily interested in how well instruments perform. That is, how closely does the instrument's output represent its input? Before beginning a discussion of performance, we should consider what functions an instrument may perform in making a measurement. We will not consider the mechanism of any particular instrument. Rather, we will examine the functions of the instrument.

The objectives of this chapter, and the sections which correspond to each of them are listed below. You should be able to accomplish each of the tasks described in the objectives.

After studying this chapter you should be able to:

1. Name and describe each functional element. (1-1)
2. Identify the functional elements present in any given instrument. (1-1)
3. Distinguish between digital and analog signals. (1-2)
4. Distinguish between instruments with deflection outputs and those with null outputs. (1-2)
5. Distinguish between active and passive transducers. (1-2)

In the introduction, we indicated that we would be interested in the performance of instruments. Before concerning ourselves with how well an instrument performs, we should examine *how* it performs its function. We do this by defining the functions of the various components of an instrument system.

The purpose of an instrument is to furnish information about the physical quantity to which it is sensitive. The term *signal* will be used to denote the carrier of this information. The transmission of information from one system to another requires the exchange of energy or power between the systems. Energy is a function of a potential and a displacement, and power is given in terms of a potential and a velocity or flow. The signal that transmits information may be the potential or the displacement (or the flow in the case of power exchange).

1-1
Functional
Elements

Consider first the minimum functional requirements of an instrument. What is the least an instrument can do and still fulfill its purpose? (Try to answer my questions before you read on. It will help you understand the material.) First, the instrument must be sensitive to the variable we are attempting to measure. Second, it must display this information in a form which can be detected by one of the *human senses*, usually sight or hearing. (In the case of control systems, this display should be in a form which the system can detect.) Do you agree? Good!

Let us call the portion of the instrument that senses the measured variable the *primary sensing element*. The section of the instrument that provides the output signal will be called the *data presentation element*.

Now consider a simple instrument, the liquid-in-glass thermometer. What portion of the thermometer would you designate as the primary sensing element? The bulb? Right! How is the data (temperature in this case) presented? By means of the fluid column and the etched scale? That is correct. Thus, in the case of the liquid-in-glass thermometer, the bulb is the primary sensing element, and the liquid column and the scale comprise the data presentation element.

Before putting the thermometer aside, consider this. Does this instrument perform any function other than sensing and display? Note that the information (or signal) is received in the form of thermal energy (heat is transferred to the bulb), but mechanical energy (displacement of the fluid column) is involved in the display function. It appears that the thermometer is capable of changing the form of the signal by which information is transmitted without altering the information. We shall call the portion of an instrument which converts the signal from one form to another the *variable conversion element*. Many conversions are possible. Some of these are shown in Table 1-1.

TABLE 1-1   Variable Conversions

Input	Output
Rotation	Translation
Displacement	Voltage Current Force
Temperature	Displacement Voltage Resistance
Pressure	Displacement Force
Voltage	Current Frequency

Now let us consider the odometer of an automobile. This instrument detects rotation of some component of the drive train and converts it to distance traveled. In order to give the desired output, a mathematical operation must be performed upon the input. In this case, the input is multiplied by a constant:

$$\text{Input} \quad \times \quad \text{Constant} \quad = \quad \text{Output}$$

$$[\text{revolutions}] \quad \left[\frac{\text{miles}}{\text{revolution}}\right] \quad [\text{miles}]$$

We will agree to call that portion of an instrument which operates upon the signal according to some mathematical rule, without changing the physical form of the signal, the *variable manipulation element*.

Is there any other new function that the odometer performs? In answering, consider the location of the primary sensing element (on the transmission) and the data presentation element (on the dashboard). How did the signal travel this distance? By means of the speedometer cable. Right! In other words, the function of the speedometer cable is to transmit the signal from the transmission to the dashboard. Let us name any element of an instrument that performs this function a *data transmission element*.

Perhaps we should take a moment to summarize the preceding paragraphs. We have defined five elements of instruments according to their functions. Identify each of the terms defined below.

1. The element that is sensitive to the measured variable.
2. The element that converts the signal from one physical form into another without changing the information content of the signal.
3. The element that operates on the signal according to some mathematical rule without changing the physical form of the signal.
4. The element that transmits the signal from one location to another without changing its information content.
5. The element that provides a display of the output in some form that can be detected by one or more of the human senses.

The following exercises are intended to give you some experience in identifying the various functional elements of instruments. In solving them, you may note that not all instruments contain all elements. Also, some components of an instrument may perform more than one function. This characteristic might make complete identification of the function rather difficult.

PROBLEMS

1-1   One method of indicating the level of water in a tank is shown in Fig. P1-1. Identify the various functional elements.

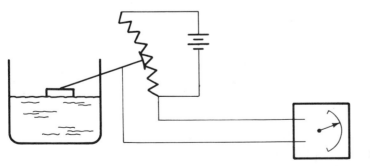

Figure P1-1.

1-2 Identify the functional elements of the dial indicator shown in Fig. P1-2.

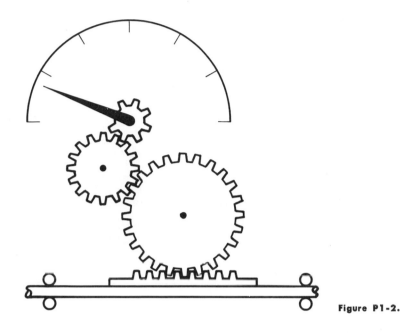

Figure P1-2.

1-3 Identify the functional elements of the aneroid barometer shown in Fig. P1-3.

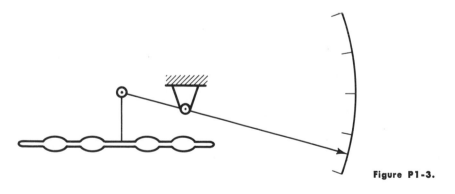

Figure P1-3.

1-4 A stereo set might be considered an instrument. The input is the displacement of the stylus as it follows the undulations of the groove in the record. The output is the sound (music?) from the speakers. Identify the various functional elements of the stereo.

1-2
Operating
Characteristics

Certain operating characteristics are also of interest to us. These characteristics may appear in any of the functional elements of a measuring system. Three important questions should be considered when examining an element:

Is the signal digital or analogous in form?
Is the output represented by a deflection or is it obtained by a null?
What is the source of power that is required for operation of the element?

We shall examine each of these.

Digital and analog signals

A signal that is *analogous* to an input is a continuous function of the input. Since it is continuous, it has an infinite number of values. Each of these corresponds to a particular value of input. For instance, the deflection of the pointer of a meter and the pen of a recorder are analog signals.

In contrast, some signals are discrete. They have a finite number of values, usually represented as *digits* in some numbering system (base 10, 2, etc.). Since the relationship between the input and output signal is not continuous, each value of the output signal may correspond to

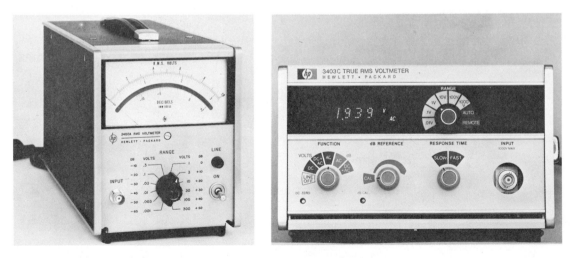

Figure 1-1. Voltmeters with (a) Analog and (b) Digital Displays. (*Courtesy Hewlett-Packard Co.*)

several values of input. Figure 1-1 shows examples of voltmeters with analog and digital outputs. It is important that the form of signal (digital or analog) generated by a particular element be compatible with the rest of the system. For instance, consider a control system containing a digital computer. Information at the input and output of the computer is in digital form. Thus, the element that provides input to the computer must provide a digital output.

Deflection and null outputs

Consider the operation of a chemical balance, Fig. 1-2. An unknown quantity that is to be weighed is placed in one pan of the balance. Weights of known value are placed on the other pan until they equal the unknown weight. This balanced condition is usually indicated by the zero (*null*) position of a pointer. Your bathroom scale is an example of an instrument with a *deflection* type of output. The weight of an object that you place on the platform of the scale is indicated by the relative displacement between a pointer and a dial.

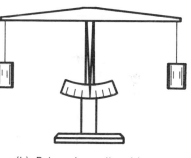

(a) Unbalanced condition   (b) Balanced or null position

Figure 1-2. The Chemical Balance

Both means of indication have advantages. In general, the null devices are more accurate (more closely indicate the variable measured), but they are unable to measure variables which change with time. Instruments that utilize deflection outputs are usually capable of measuring time-dependent signals. However, they are not as accurate as null devices.

Passive and active devices

Do you recall what was involved in the process of transmitting information from one system to another? If not, stop now and review the material on the first page of this chapter.

An energy transfer must occur when information is transmitted from one system to another. This energy transfer will often cause the variable to change as it is being measured. To minimize this change, we must consider the power required to operate an instrument and the source of the power.

Consider a primary sensing element that receives energy (or power) from some system and transmits energy to the next element of the instrument. In some cases all of the energy transferred from the primary sensing element is obtained from the system upon which the measurement is performed. We refer to elements that operate in this fashion as *passive elements*. Since we usually wish to minimize the energy removed from the measured system, the output of passive elements is usually low.

Often an auxiliary source of energy is introduced so that the output of the element can be increased without significantly increasing the energy removed from the measured system. The electronic amplifier is one of the most common examples of this kind of element. We call elements that utilize an auxiliary source of energy *active elements*.

PROBLEMS

1-5 The following is a list of devices that should be familiar to you. State whether the output from the device has a digital or an analog type output.

(a) The timer on a football scoreboard

(b) Your wristwatch

(c) The speedometer of your automobile

(d) The fuel gage of your automobile

(e) The calibrated balance beam of a platform scale

(f) The odometer of your automobile

1-6 Which of the following devices have null outputs? Which have deflection outputs?

(a) A spring scale (such as your bathroom scale)

(b) A platform scale (such as used in a physician's office)

(c) The fuel gage of your automobile

(d) The light meter in a camera

(e) A thermometer

(f) Your wristwatch

1-7 Which of the following devices can be classified as passive? Which are active?

(a) A vacuum tube voltmeter

(b) The speedometer of your car

(c) The pickup in the tone arm of a record changer

(d) The fuel gage of your automobile

(e) A thermometer

(f) A light meter for a camera

1-3
Some Basic
Instruments
Now that we have considered how instruments operate in general, let us apply these ideas to some specific instruments. In the following sections you will be introduced to a number of the most commonly used instruments. In this chapter we will concentrate on how each instrument operates. We will discuss the performance of these instruments in later chapters.

Displacement

We often wish to measure the position of one object relative to another or the displacement of an object with respect to some reference point. There are a number of devices which are used to sense this displacement. We will consider only three, the dial indicator, the strain gage, and the linear variable differential transformer (LVDT).

You encountered the dial indicator in Prob. 1-2. The linear motion of the spindle is translated into an angular displacement of the instrument pointer by means of the gear train shown. By proper selection of the gear ratios, small displacements of the spindle will result in a relatively large motion of the pointer. The range of these inputs varies from displacements of zero to 12 in. for typical dial indicators. The smallest displacement that can be measured is as small as 0.00005 in.

Now, let us consider the functional elements of the instrument. The spindle is the portion of the instrument that is sensitive to the input displacement. You will recall that the part of an instrument that performs this function is called the *primary sensing element*. What are the functions of the gear train? There are several. First, the gears transmit the input signal from the spindle to the pointer. They also alter the form of the signal from translation to rotation. Finally, the gears amplify (multiply) the input so that a large output displacement occurs. Referring to section 1-1, we see that the gear train serves as a *variable transmission element*, a *variable manipulation element*, and a *variable conversion element*. The pointer and associated scale comprise the *data presentation element*.

If we consider the operating characteristics of the dial indicator, we see that it has an *analog* output and is a *deflection* type device. Since all of the energy required to operate the instrument is furnished by the system whose displacement is being sensed, we classify the dial indicator as a *passive* device.

A second and very widely used displacement sensing device is the strain gage. Because of the great utility of this instrument, we will describe it in some detail. In 1856 Lord Kelvin discovered that the resistance of electrical conductors changes when they are subjected to mechanical strain. It was not until 1938 that this phenomenon was applied to produce the strain gage as we now know it. You will usually find the strain gage in one of two forms: *bonded gages*, which are firmly cemented to a surface over their entire length, and *unbonded gages*, which are connected only at the ends of the gage. Let us look at the bonded type of gages first.

Figure 1-3 shows a typical foil gage. It is composed of a metallic grid, which is usually etched from a thin foil of metal. This grid is mounted on a thin film of mylar, or similar material, to provide both mechanical strength and electrical insulation. When the gage is to be

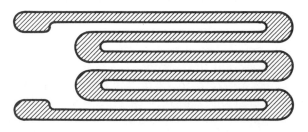

Figure 1-3. A Typical Foil Strain Gage

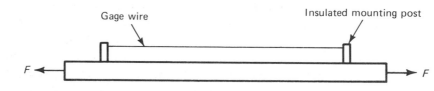

Figure 1-4. A Simple Unbonded Strain Gage

used, the assembly is cemented in place at the location where strain is to be measured. We will not concern ourselves with the details of this process except to note that, in order for the gage to function properly, the bond must be strong enough to prevent motion between the gage and surface.

In its simplified form, the unbonded gage is simply a wire which is stretched between two points as shown in Fig. 1-4. Note that the wire must be under some initial tension if both positive and negative displacements are to be sensed. For reasons that will be discussed shortly, the unbonded gage usually consists of four wires. A simple arrangement of this sort is shown in Fig. 1-5. The unbounded gage is usually

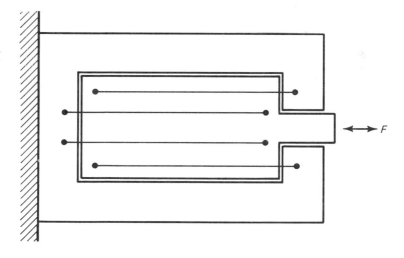

Figure 1-5. A Simple Four-Wire Unbonded Strain Gage

found as a component of some other instrument. One or two possible applications will be pointed out in a later section.

It is possible to derive a theoretical relationship between resistance change and strain, but this is not particularly useful to us. Rather, we use the following empirical (experimentally determined) relationship:

$$\frac{\Delta R}{R} = F\frac{\Delta L}{L} \tag{1-1}$$

R = resistance of the gage when not strained, (ohms)
L = length of the unstrained gage, (inches)
ΔL = change in length of gage when under strain, (inches)
ΔR = change of resistance because of ΔL, (ohms)
F = gage factor, (an experimentally determined constant)

As you may know, $\Delta L/L$ is the unit strain and is usually given the symbol, ϵ. Thus, Eq. (1-1) is usually written

$$\frac{\Delta R}{R} = F\epsilon \tag{1-2}$$

Most of the commonly used strain gages have gage factors which range between 2.0 and 3.5. Usual values of the nominal resistance, R, are: 120, 350, and 500 ohms. The maximum allowable strain may range from 0.005 to 0.20 in. per inch. From this we can see that the change in resistance is usually quite small. For instance, for a 120 ohm gage with a gage factor of 2.0 and a strain of $1,500 \times 10^{-6}$ in. per inch,

$$\Delta R = RF\epsilon = 120(2 \times 1500 \times 10^{-6}) = 0.36 \text{ ohm}$$

The fact that ΔR is so small can cause some difficulty in measuring it. To avoid this problem, we usually use a Wheatstone bridge circuit. This very useful circuit is shown in Fig. 1-6. Any, or all, of the four resistances shown may be a strain gage. Let us examine the Wheatstone bridge in some detail.

If we apply the techniques of circuit analysis to the bridge, we find the output, e_o, is zero when

$$\frac{R_1}{R_2} = \frac{R_3}{R_4} \tag{1-3}$$

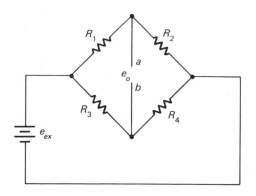

Figure 1-6. The Wheatstone Resistance Bridge

The bridge is said to be *balanced* when the output is zero. If one or more of the resistances are changed so that Eq. (1-3) is not satisfied, e_o will have some value. We can use this to our advantage in one of two ways. One possibility is to alter the circuit by adding a variable resistor as shown in Fig. 1-7. Let us suppose that R_1 represents a strain gage

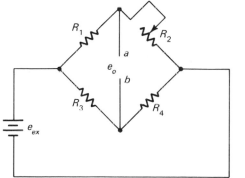

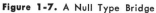

Figure 1-7. A Null Type Bridge

with a resistance of 120 ohms and that R_3 and R_4 are fixed resistors of 120 ohms each. The variable resistor, R_2 is initially set at 120 ohms. If a strain is imposed on the gage, R_1 will change to some value, thus causing the bridge to become unbalanced. We can rebalance the bridge and make e_o zero once again, by introducing some additional resistance in R_2, in other words, by adjusting the variable resistor. From Eq. (1-3), we can see that the resistance that must be added is equal to the change of resistance of the strain gage. Thus, it is possible to calibrate the variable resistance in terms of strain. Since the output of the bridge is always restored to zero, you should recognize that this is a null output device.

It is also possible to use the Wheatstone bridge circuit as a deflection output device. In this arrangement, we initially adjust the resistances so that the bridge is balanced. After that, we allow the bridge to stay unbalanced as the gages are strained and observe the output. Refer to Fig. 1-6 and assume that R_1 is a strain gage. If a strain is applied, the resistance will increase by ΔR. Analysis of the circuit will show that the output voltage is

$$\frac{e_o}{e_{ex}} = \frac{\Delta R}{4R}$$

if there is no current between points a and b. Let us now assume that R_2 is the strain gage and that its resistance increases by ΔR. In this case, circuit analysis will show us that

$$\frac{e_o}{e_{ex}} = \frac{-\Delta R}{4R}$$

Be sure to remember this. You will probably be able to use it later.

Now, what happens when more than one of the resistances in the bridge is a strain gage? That, of course, depends on the way the resistance changes. Referring to Fig. 1-6 again, if R_1 and R_4 increase by ΔR, and R_2 and R_3 decrease by ΔR

$$\frac{e_o}{e_{ex}} = \frac{\Delta R}{4R_1} + \frac{\Delta R}{4R_4} - \frac{(-\Delta R)}{4R_2} - \frac{(-\Delta R)}{4R_3} \qquad (1\text{-}4)$$

if

$$R_1 = R_2 = R_3 = R_4 = R$$

we can write

$$\frac{e_o}{e_{ex}} = \frac{4\Delta R}{4R}$$

Or, in general,

$$\frac{e_o}{e_{ex}} = \frac{n\Delta R}{4R} \qquad (1\text{-}5)$$

where n is the number of arms of the bridge that contain strain gages. *Be careful when applying this equation.* The resistance changes must be as indicated in Eq. (1-4) or Eq. (1-5) is not valid. We can write Eq. (1-5) in terms of strain by means of Eq. (1-2). Thus,

$$\frac{e_o}{e_{ex}} = \frac{n}{4} F\epsilon \qquad (1\text{-}6)$$

You might be wondering how to determine how many of the arms in the bridge should (or could) be *active*, that is, how many contain strain gages. There is no general rule but the following pointers may be helpful. Look first at Eq. (1-6). Note that e_o/e_{ex} is largest for $n = 4$. We usually want the output to be as large as possible so this is the configuration we would usually choose, if possible. Are there any applications in which this is not possible? Look at Eq. (1-4). Notice that two of the arms must experience positive strain (ΔR positive) whereas the other two experience negative strain (ΔR negative). This is always the case if an unbonded bridge is involved. In many applications where bonded gages are used, however, this is not possible. In those cases one or two arms are active. Figure 1-8 shows several applications in which the bridge contains fewer than four active arms.

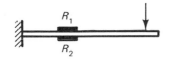
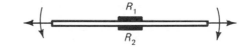

(a) Gages will sense bending strain

(b) Gages will sense bending strains, but not tensile strains

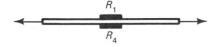
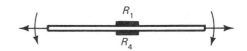

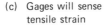

(c) Gages will sense tensile strain

(d) Gages will sense tensile strains, but not bending strains

Figure 1-8. Strain Gage Applications with Two Active Gages

PROBLEM

1-8   Refer to Fig. P1-8. Identify the functional elements that are present. Is the bridge a null or deflection type device? Is the output analog or digital? Is the strain gage an active or passive device?

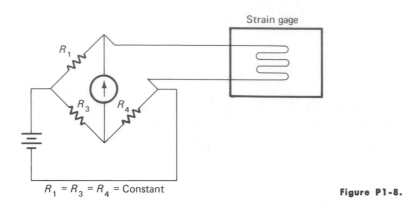

$R_1 = R_3 = R_4 =$ Constant

Figure P1-8.

The last displacement sensing device that we will consider is the linear variable differential transformer (LVDT). You have probably encountered transformers in some form or other. Maybe you had an electric train or a race set when you were younger. Although the function of the LVDT is different from that of your train transformer, the basic principles are the same. A simple transformer is shown in Fig. 1-9. If we impose an alternating (AC) voltage on the primary winding,

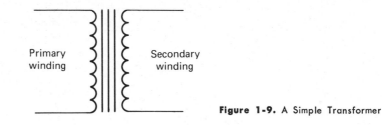

Primary winding   Secondary winding

Figure 1-9. A Simple Transformer

an AC voltage of the same frequency will be induced in the secondary winding. The magnitude of the induced voltage depends upon the number of turns in the windings, the geometry of the device, and the material in the core.

The LVDT is shown schematically in Fig. 1-10(a). Notice that there are two secondary windings and that the core material is air, with a movable iron rod which is inserted in the tubular core. Each secondary winding will have an induced voltage, which has the same frequency as the primary voltage. The magnitude of the induced voltages depends upon the position of the iron rod. Figure 1-10(b) shows how the secondary voltage changes with the position of the iron rod.

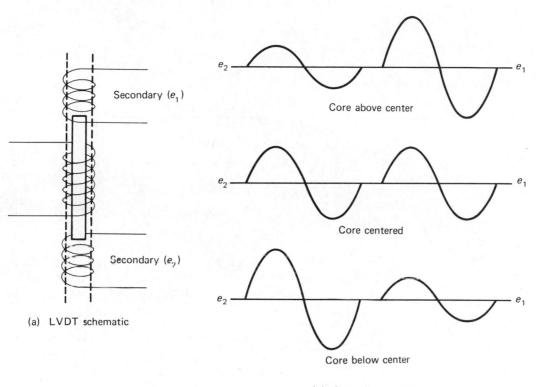

Secondary (e_1)

Secondary ($e_{\prime\prime}$)

(a) LVDT schematic

e_2 — — — — e_1

Core above center

e_2 — — — — e_1

Core centered

e_2 — — — — e_1

Core below center

(b) Secondary voltages

Figure 1-10.

You can probably see how we can use this arrangement to measure displacement. If the iron rod is connected to the part whose displacement is to be measured, each of the secondary voltages is a function of position. When the secondaries are connected in series as shown in Fig. 1-11(a), the output is proportional to position for some portion of core travel as shown in Fig. 1-11(b). Although the output of the secondaries is an AC voltage, we can rectify and filter this signal to obtain a DC output that is proportional to displacement. We will not examine this circuitry.

The range of displacement that can be sensed by LVDT's may be as high as 10 in. The minimum displacement that can be sensed is about 1×10^{-6} in.

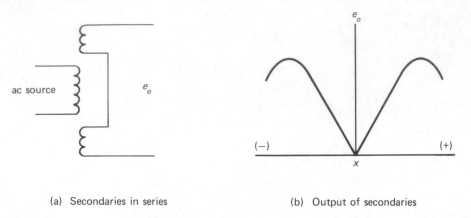

(a) Secondaries in series   (b) Output of secondaries

Figure 1-11.

PROBLEM

1-9   Refer to the LVDT shown in Fig. P1-9. Identify each of the functional elements present. Is the LVDT a null or deflection type of device? Is the output analog or digital? Is the LVDT an active or passive device?

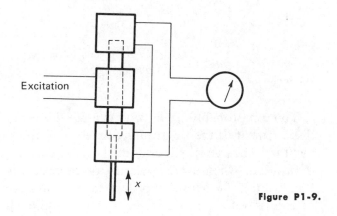

Figure P1-9.

Velocity and acceleration

The devices we will discuss in this section are designed to measure absolute velocity and acceleration, that is, the velocity or acceleration with respect to the surface of the earth. There are several ways in which these measurements can be made. However, we will restrict our examination to a single class of instrument—the seismic accelerometer. This may seem restrictive, but most of the required velocity and acceleration measurements can be accomplished with this type of instrument.

How can we use an accelerometer to measure velocity? Let us look at the definitions of acceleration and velocity. The acceleration, a, is defined as the second derivative of displacement with respect to time.

$$a = \frac{d^2x}{dt^2}$$

Velocity is the first derivative of displacement with respect to time.

$$v = \frac{dx}{dt}$$

Now, can you see the relationship between acceleration and velocity?

$$a = \frac{dv}{dt}$$

or

$$v = \int a dt \tag{1-7}$$

Since it is relatively easy to integrate an electrical signal, we have no particular difficulty in measuring velocity using an accelerometer.

Now, let us see how a typical accelerometer operates. We can do this by means of a simple model, Fig. 1-12. Note that the components of the accelerometer are the frame, which is rigidly attached to the device whose acceleration we wish to measure, a mass, which is supported by a spring, and a damper. The function of the damper will be

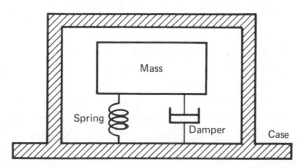

Figure 1-12. Model of an Accelerometer

discussed in later chapters. Assume that a constant acceleration is imposed on the frame. What is going to happen to the mass? It, also, will accelerate. We know that the mass will not be accelerated unless some force is applied (remember $F = ma$?). What is the source of this force? It is transmitted from the frame to the mass by the spring. The spring is deformed in proportion to the force that it transmits. Thus the mass moves relative to the frame when an acceleration is applied. Further, the motion is proportional to the acceleration. The final element we need is a means of sensing this displacement. Any one of the displacement-sensing devices we discussed would be suitable for this purpose. The unbonded strain gage transducer is particularly well suited for this application.

The range of accelerometers is usually stated as some multiple of the acceleration of gravity. Thus, 10 g means 10 times the acceleration of gravity or about 322 ft/sec². We can obtain accelerometers that have a range of ± 0.25 g through $\pm 10,000$ g.

Force

There are a number of ways to sense force. We will consider only two of these. You have probably encountered both of these in some form. First, we can compare the unknown force with the weights of objects of known mass. The analytical balance (Fig. 1-1) is an example of this type of device. Fig. 1-13 shows a platform scale that also operates in this way. Note that the system of levers provides mechanical advantage so that the balance weight is much smaller than the force that is to be measured. The linkage arrangement also allows the force to be measured accurately, regardless of the exact point of its application on the platform.

A second way to detect force is to apply the force to some elastic member and observe the deformation of the member. The spring scale shown in Fig. 1-14 is an example of this kind of instrument. The elongation of the spring is directly proportional to the applied force. A scale, calibrated in pounds, is generally attached to the case of the device to allow direct measurement of the force.

We can also apply the principle of elastic deformation in other ways. For instance, a small metallic cylinder, as shown in Fig. 1-15, can be used. When a force is applied as shown in the figure, the height of the cylinder will change. This deformation is proportional to the force

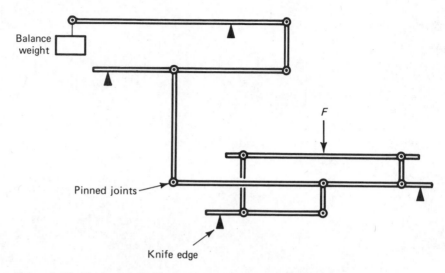

Figure 1-13. Platform Scale Mechanism

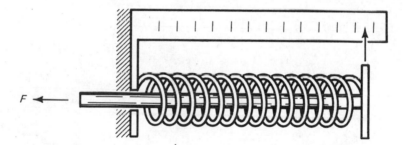

Figure 1-14. A Spring Scale

applied. We can detect this deformation in several ways. One of the most common is to bond strain gages to the side of a cylinder. Several other elastic elements that might be used are also shown in Fig. 1-15. Figure 1-16 contains several elastic elements that are used to convert force (the ring and slotted cylinder) or pressure (the other devices) into displacement. Devices of the type we have been discussing are called *load cells*.

Load cells are available in a wide variety of ranges. One of the smallest, which uses an unbonded strain gage arrangement, has a range of ± 3.0 gm. One of the largest of these devices which uses a bonded strain gage, is capable of sensing forces up to 250,000 lbs.

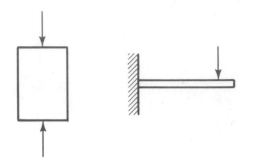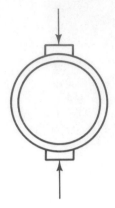

Figure 1-15. Elastic Elements for Load Cells

Figure 1-16. Elastic Elements that Convert Force and Pressure to Displacement. (*Courtesy Schaevitz Engineering*)

PROBLEM

1-10   Consider Fig. P1-10. Identify the functional elements that are present. Is this a null or deflection type of device? Is the output digital or analog? Is this an active or passive device?

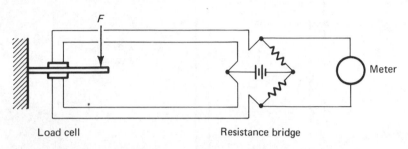

Load cell   Resistance bridge

Figure P1-10.

Pressure

Generally, we measure pressure by transducing it to a force, applying the force to an elastic element, and observing the resulting deflection. Figure 1-16 shows examples of these elastic elements. There are other useful techniques that we will not cover here. These other methods are usually used to measure extremely high or low pressures.

One of the simplest and most useful pressure measuring devices is the manometer. In its most elementary form, it takes the form of a U-shaped tube which is partly filled with some liquid as shown in Fig. 1-17. As in the case of most other pressure gages, the manometer senses pressure differences. If we apply the principles of fluid statics to the manometer, we find that

$$P_B - P_A = \gamma h \tag{1-8}$$

if the fluid is in static equilibrium.

There are several variations of the manometer. For instance, if we make the diameter of one of the legs much larger than that of the other, we have what is termed a *well*, or *cistern*, type of manometer (Fig. 1-18). Note that the elevation of the fluid in the large diameter leg will change very little compared to that in the small leg. Generally, we neglect the large leg and consider *h* in Eq. (1-8) to be the length of the fluid column in the smaller leg.

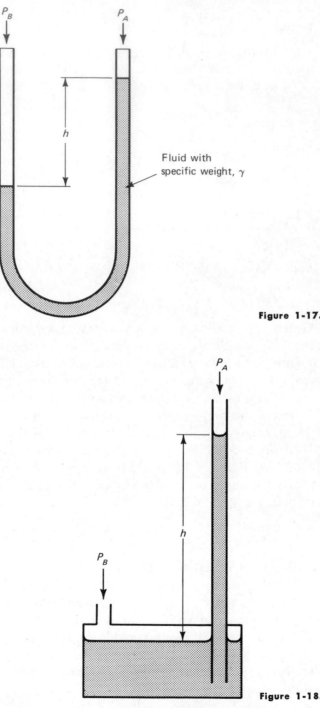

Figure 1-17. U-Tube Manometer

Fluid with specific weight, γ

Figure 1-18. Well-Type Manometer

26

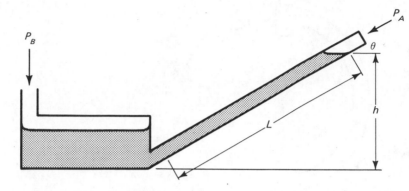

Figure 1-19. A Draft Gage

Another variation is obtained by inclining the small leg of a cistern manometer as shown in Fig. 1-19. The pressure difference is still given by Eq. (1-8), but notice how far the fluid moves in the tube for a given pressure difference. If we call this distance L, we see that

$$L = \frac{h}{\sin \theta} \tag{1-9}$$

If, for instance, $\theta = 5°$, and $h = 1$ in.

$$L = 1/.0872 = 11.47 \text{ in.}$$

In constructing a manometer of this type (usually called a *draft gage*) we usually place the scale along the tube and expand it according to Eq. (1-9), so that the scale actually indicates the vertical displacement of the fluid, h.

The range of pressures that can be measured with manometers depends upon the fluid used, the minimum displacement which can be sensed, and the tube length. The fluids used are generally water, alcohol, oil, or mercury. Pressures as low as 0.00005 psi can be measured using these devices. It is not unusual to measure pressure differences in the range of 20–30 psi using manometers.

Another pressure sensitive device that is widely used is the bourdon tube gage. This is a tube with a more or less oval cross section, which is formed as shown in Fig. 1-20. Pressure applied to the inside of the tube will cause the cross section to become more rounded. This will make the tube straighten out. The displacement of the end of the tube

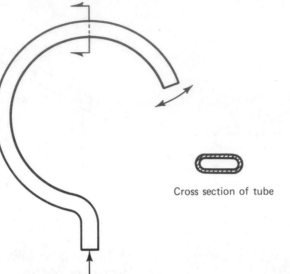

Cross section of tube

Figure 1-20. A Bourdon Tube

is proportional to pressure. We can sense this displacement in a number of ways. Very often, the end of the tube is connected to a pointer by a suitable gear train.

Bourdon tube gages have a much wider range than manometers. Depending upon the material used, pressures in excess of 10,000 psi can be measured.

The third type of gage we will discuss is the diaphragm gage. The elastic element in this gage is a thin disc that is firmly clamped around its edge (Fig. 1-21). If we increase the pressure on one side of the diaphragm, it will deflect as shown by the broken lines in the figure. This deformation, which is not quite directly proportional to pressure, can be sensed in either of two ways. We can measure the deflection at the center of the gage with a suitable instrument, or we can bond strain gages to the surface of the diaphragm. Both methods are used in commercial gages.

This type of gage is available in about the same ranges as bourdon tube gages.

Flow

We can utilize several physical phenomena to sense the rate of flow of a fluid. For instance, if we are measuring the flow of a liquid, we can measure the length of time required to collect a given volume or weight of liquid in a container. Or we might measure the average

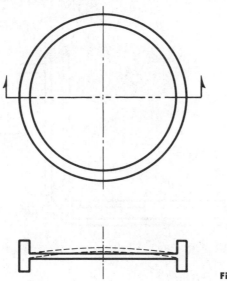

Figure 1-21. Pressure Diaphragm

velocity of the fluid and multiply the result by the area through which the fluid is flowing. A third method involves partially obstructing the flow area and measuring the resulting pressure drop. As a fourth alternative, we might observe the drag forces due to the flow. Finally, we might use part of the energy in the fluid to drive a turbine or motor. Variations of all of these methods are found in commercial flow meters.

Although it may seem rather crude, the method of collecting a measured volume or weight of liquid can give very accurate measurements. Often this technique is used to calibrate other flowmeters. Our ability to measure the time elapsed and volume collected determine the magnitude of the errors which occur.

The instrument that is most often used to measure fluid velocity is the pitot-static tube. This instrument is illustrated in Fig. 1-22. We can use our knowledge of fluid mechanics to learn how the pitot tube operates. Imagine that we can identify a stream of fluid flowing from the point numbered 1 right into the end of the pitot tube. Since the other end of the tube is connected to some pressure sensing device, the velocity of the fluid stream becomes zero at the end of the tube. Application of the Bernoulli equation gives

$$\frac{P_1}{\gamma} + \frac{V_1^2}{2g} = \frac{P_0}{\gamma}$$

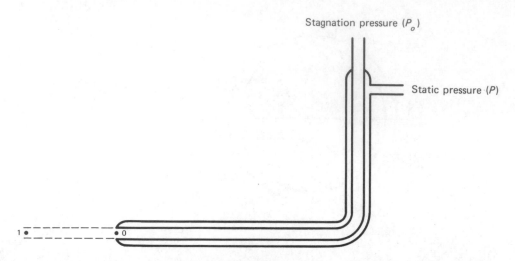

Stagnation pressure (P_o)

Static pressure (P)

Figure 1-22. A Pitot-static Tube

where

$P =$ pressure, pounds per sq ft
$\gamma =$ the specific weight of the fluid, pounds per cu ft
$V =$ the fluid velocity, ft/sec, and
$g =$ the acceleration of gravity.

When we solve this equation for velocity, we find that

$$V_{\text{ideal}} = \sqrt{2g\frac{(P_0 - P_1)}{\gamma}} \tag{1-10}$$

In practice, we find that Eq. (1-10) gives velocities that are too high. This occurs because the real physical situation is slightly different from that required by the Bernoulli equation. We can correct this by multiplying Eq. (1-10) by an experimentally determined coefficient C. Thus,

$$V_{\text{actual}} = C\sqrt{2g\frac{(P_0 - P_1)}{\gamma}} \tag{1-11}$$

The value of C is usually between 0.98 and 1.02.

It is often inconvenient to measure P_1 at the point indicated in Fig. 1-20. Fortunately, it is possible to locate a point on the outer wall of the pitot-static tube where the pressure is equal to P_1. If holes are drilled at this point, as shown in Fig. 1-20, both P_1 and P_0 can be sensed by the instrument.

How can we obtain flow rate from velocity? Suppose we have air flowing in a rectangular duct as shown in Fig. 1-23. Let us divide the

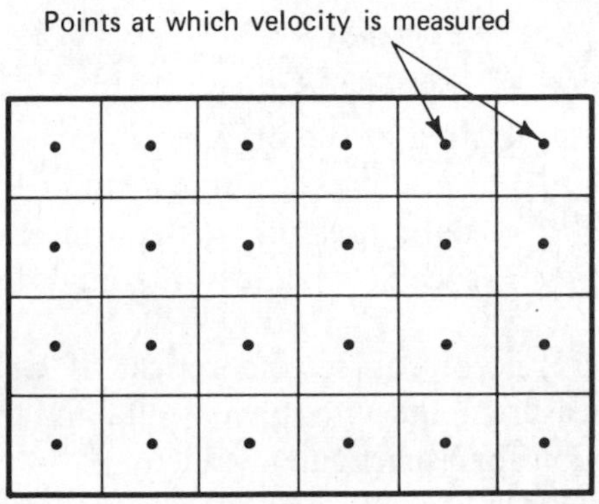

Points at which velocity is measured

Figure 1-23. Duct Divided into Smaller Areas

cross section of the duct into a number of smaller areas and measure the velocity at the center of each of these. If the areas are small enough, we can assume that the velocity at the center is the average velocity for that area. Since the volumetric flow rate, Q, is the product of the average velocity and the cross sectional area,

$$Q = V_1 A_1 + V_2 A_2 + \cdots + V_n A_n$$

This idea can be extended to a flow area of any shape.

Now, let us consider what happens if we partially obstruct the flow in a duct. Figure 1-24 shows a sharp-edged orifice placed in a duct. This is simply a thin plate with a hole that is concentric with the duct.

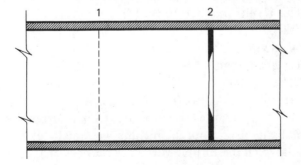

Figure 1-24. Orifice Meter

If we apply the Bernoulli equation and the continuity equation to this arrangement, we find that

$$Q = \frac{A_2}{\sqrt{1 - (A_2/A_1)^2}} \sqrt{\frac{2g}{\gamma}(P_1 - P_2)} \tag{1-12}$$

where

$A_1 =$ area of the duct
$A_2 =$ area of the orifice
$P_1 =$ static pressure upstream of the orifice
$P_2 =$ static pressure at the orifice, and
$\gamma =$ specific weight of the fluid flowing.

Equation (1-12) is, of course, theoretical. If we built an orifice and made flow measurements with it, we would find that the flow predicted by this equation for a particular value of $P_1 - P_2$ is greater than the actual flow. This is because of the losses that occur in the device that are not accounted for in Eq. (1-12) and because we cannot measure the static pressure exactly at the location of the orifice.

We can correct Eq. (1-12) by introducing a discharge coefficient C. The value of the discharge coefficient varies slightly with the flow rate. In most cases, however, we can consider its value to be constant over a wide range of flow rate. The value of C_D is usually between 0.6 and 0.8.

Figure 1-25 shows one method which uses the drag force on a body to measure the fluid flow rate. If we consider the forces on the float, we find that they consist of the weight of the float, the buoyant force on the float, and the drag force on the float due to the velocity of the fluid flowing past it. The first of these forces is directed downward whereas the others are directed upward. Since the drag force is dependent upon velocity, an increase in flow will cause the drag force to increase. This results in a net upward force that will cause the float to rise in the tube.

Note that the tube is tapered. Upward motion of the float causes a decrease in velocity and a restoration of equilibrium of the forces on the float. The float has a unique position for each value of flow.

Finally, let us consider a class of flowmeters in which some of the energy of the fluid stream is used to turn a small turbine. Rather than attempt to obtain work from the turbine, we simply let it rotate freely. The speed at which the turbine rotates is proportional to the flow rate of the fluid passing through the meter.

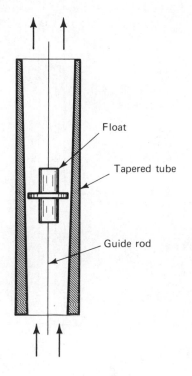

Float

Tapered tube

Guide rod

Figure 1-25. A Drag Force Meter

Temperature

We will discuss only two of the many methods available to sense temperature. These are the liquid-in-glass thermometer and the thermocouple. You are more likely to encounter these two devices than any others.

The liquid-in-glass thermometer is one of several devices that utilize the fact that the dimensions of liquids and solids depend upon temperature. In the case of the liquid-in-glass thermometer, a quantity of liquid (usually mercury or alcohol) is contained in a bulb that is connected to a small diameter tube (Figure 1-26). Changes in temperature will cause the volume of the fluid to change. Since the area of the tube is much less than the bulb, the relatively small changes of fluid volume will result in a significant change in the length of the filament. The freezing and boiling points of the liquid used determine the minimum and maximum temperatures that can be measured. For a mercury filled thermometer, the range is $-38°$–$950°F$. The range for an alcohol thermometer is $-95°$–$150°F$.

Thermocouples take advantage of a different physical phenomenon.

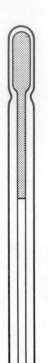

Figure 1-26. Liquid-in-glass Thermometer

The thermocouple is composed of two wires, of different materials, which are connected together at junctions as shown in Fig. 1-27. If the junctions are at different temperatures, a current will flow in the wires. This indicates that the junctions are at different electrical potentials. We can use this phenomenon to sense the difference between the junction temperatures.

Before we look at methods of sensing this potential difference, let us study one additional concept. Suppose we add a third metal to our circuit as shown in Fig. 1-28. If the temperature of this wire is uniform over its whole length, the potential between the junctions *A* and *B* is unaffected. This principle is referred to as the *law of intermediate*

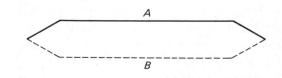

Figure 1-27. Thermocouple Circuit

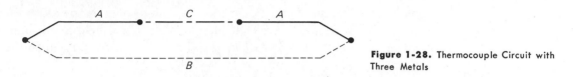

Figure 1-28. Thermocouple Circuit with Three Metals

metals. As we shall see, this characteristic is extremely valuable to us in using the thermocouple to make temperature measurements.

We now have a device that produces an electrical potential that is a function of temperature difference. How can we measure this potential and thereby find the temperature? Suppose we place an ammeter or voltmeter in the circuit, as shown in Fig. 1-29. Note that the terminals

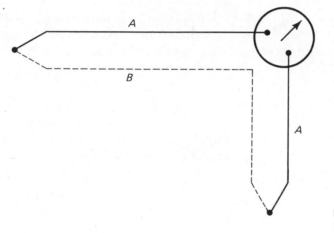

Figure 1-29. Thermocouple with Meter in Circuit

of the meter are probably composed of a different material from that of the thermocouple wire. Does this have an effect on the potential we are trying to sense? According to the law of intermediate metals, it does not, as long as the meter is at a uniform temperature.

We can now measure the *difference* between the junction temperatures. What if we wish to know the temperature at one of the junctions? We can determine this only if we can determine the temperature of the other junction in some way. This can be done either by measuring the junction temperature with some other temperature-measuring device or by placing the junction in a place where the temperature is known. Can you think of a simple way to generate a known temperature? What about a mixture of ice and water? We know that the temperature of a mixture of ice and water is 32°F. This is very easily obtained and is quite often used to provide a known temperature for one of the junctions. We refer to the junction that is held at constant temperature as the *reference junction*. The other junction is called the *measuring junction*.

Sometimes it appears that there is no reference junction. Refer to Fig. 1-30. Where is the reference junction in this case? We can imagine

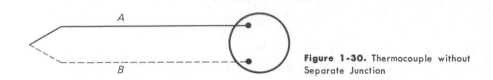

Figure 1-30. Thermocouple without Separate Junction

that the meter represents a third material placed in the circuit. According to the law of intermediate metals, we can ignore the presence of this metal if its temperature is uniform. Since this is usually the case for a meter, we can safely ignore the effect of its presence. This indicates that the reference junction is at the meter terminals.

Although virtually any pair of metals will behave as we have described, some pairs are better suited for use as thermocouples than others. Those most commonly used, and their ranges, are as follows: (note that the thermocouples are generally referred to by type rather than the names of the two metals used). *Type T* (Copper-Constantan), $-300°-+500°$F, *Type J* (Iron-Constantan), $0°-1400°$F, *Type K* (Chromel-Alumel), $0°-2300°$F, and *Types R* and *S* (Platinum-Rhodium-Platinum), $0°-2700°$F.

PROBLEMS

1-11   Refer to Fig. 1-19. Identify the functional elements that are present.

1-12   Refer to Fig. 1-25. Identify the functional elements that are present.

1-13   Refer to Fig. 1-30. Identify the functional elements that are present.

2

Elementary

Statistics

Very often we perform an experiment several times. Although we try to perform all repetitions under identical conditions, we usually find that the results are not identical. A simple example is to determine how far a toy car will coast down its track. If you try this experiment, you will find that the car stops in a slightly different place each time. Is there some way that we can describe these results? Of course there is. Perhaps you are familiar with the techniques that we will develop in this chapter. Even if you are not, you will be able to accomplish the following tasks after you have completed your study:

1. Compute the mean value and standard deviation of a set of numbers (2-1)
2. Compute the probability that a sample from a normally distributed population will be within a specified range (2-2)
3. Determine the equation of the straight line that best fits a set of data (2-4)
4. Describe the dispersion of data about a least-squares-fit-line by computing the standard error of estimate and establishing confidence limits. (2-4)

As we will see in the next chapter, instruments have the distressing tendency to exhibit slightly different values of output for repeated applications of the same input. How can we describe such behavior? A more vital question might be, how can we predict the value of the next output? Perhaps the solution is obvious to you. We need to apply some of the concepts of probability and statistics.

**2-1
Mean Value
and Standard
Deviation**

Let us imagine that we have performed a simple experiment, which consisted of weighing sacks that are supposed to contain 100 lbs of sugar.

We can represent the results by plotting a graph showing the number of sacks that had weights within each of several ranges. The results might be as shown in Fig. 2-1. As we would hope, most of the weights are near 100 lb and the number deviating greatly from this value is relatively small.

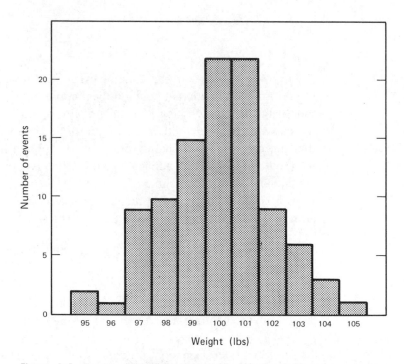

Figure 2-1. Results of Weighing One Hundred Sacks of Sugar

How can we describe results of this nature so that others can use them? We have already done one useful thing by plotting Fig. 2-1. In addition, the *average*, or *mean value*, of the numbers is of interest. This is computed by summing the results and dividing by the number of results, 100 in this case.

$$\text{mean value} = \bar{X} = \frac{\sum_{i=1}^{n} X_i}{n} \tag{2-1}$$

This is certainly useful information, but what can we do to measure the scatter or dispersion of the weights? Again, we could do the obvious and specify the difference between the maximum indication and the minimum indication. Let us call this the *range*. This, too, is useful, but it may be misleading. Consider the following set of ten numbers:

$$90, 98, 101, 96, 102, 99, 97, 108, 99, 100$$

The mean value of these numbers is ____ and the range is ____ (come on, fill in the blanks). Right! 99 and 18. Now plot a point for each reading along the line provided below.

90	95	100	105	110

What do you notice about this plot? That's right. Eight of the ten numbers are between 96 and 102. Only two, the extreme values, are out of this range of values. Thus, we see that specification of the range may give rather misleading information about the way the numbers are scattered about the mean value.

If the range is not useful, what other method of describing the scatter might we use? Suppose we decide upon specifying the difference between the mean value and each indication. We will call this value the *deviation*. Let us call the mean m, each individual indication X and the deviation D. Thus,

$$D = X_i - m \tag{2-2}$$

For our ten numbers the deviations are

X	D
90	—9
98	—1
101	+2
96	—3
102	+3
99	0
97	—2
108	+9
99	0
100	+1

This is interesting, but note that we still must present ten deviations to describe the scatter. What can we do now? What about averaging the deviation? That sounds good. We proceed by summing the ten deviations from the above table. Upon doing so, we find that the sum is zero. In fact, it is not too difficult to show that the sum of the deviations is always zero. This difficulty can be circumvented by summing the absolute value. We call this the *mean deviation* (MD).

$$\text{MD} = \sum_{i=1}^{N} \frac{|X_i - m|}{N} \qquad (2\text{-}3)$$

For our ten numbers, the mean deviation is

$$\frac{9 + 1 + 2 + 3 + 3 + 0 + 2 + 9 + 0 + 1}{10} = 3$$

Suppose we plot $m + \text{MD}$ and $m - \text{MD}$ on the line with the 10 individual indications. How about that! This appears to be (and is) a better description of the scatter than the range. (This example is perhaps a little misleading since we do not usually find 80 percent of the numbers within plus or minus one mean deviation.)

How many points could we expect to be within plus or minus one deviation of the mean value? Unfortunately, we do not know. It would be valuable to be able to make some statement of this nature. An expression which is descriptive of the dispersion of a set of measurements and (as we will see presently) allows us to state the probability that a measurement will be within a given distance from the mean value is the *standard deviation*, σ.

$$\sigma = \sqrt{\sum_{i=1}^{n} \frac{(X_i - m)^2}{N}} \qquad (2\text{-}4)$$

or, if we define

$$x_i = (X_i - m)$$

then,

$$\sigma = \sqrt{\sum_{i=1}^{N} \frac{(x_i)^2}{N}} \qquad (2\text{-}5)$$

In some instances, we may speak of the *variance*. This is simply the square of the standard deviation.

$$\text{Variance} = \sigma^2 = \frac{\sum_{i=1}^{N} (X_i - m)^2}{N} \qquad (2\text{-}6)$$

Quite often we do not know the mean value independently but we must compute it from the data at hand. In this case, the standard deviation should be computed from the following equation:

$$\sigma = \sqrt{\frac{\sum_{i=1}^{N} (x_i)^2}{N - 1}} \qquad (2\text{-}7)$$

PROBLEMS

2-1   The weight of the members of a class of 10 students is given below:

165, 208, 187, 192, 143, 156, 160, 175, 177, 168.

Compute the mean weight of the class and the standard deviation.

2-2   Fifteen ball bearings which were selected at random have the following diameters:

0.501, 0.495, 0.498, 0.503, 0.496,
0.499, 0.500, 0.506, 0.497, 0.498,
0.501, 0.502, 0.501, 0.496, 0.507.

Compute the mean diameter of the balls and the standard deviation.

**2-2
Probability**   How can we use the standard deviation to describe the dispersion or scatter of our data? As a first step, let us continue our experiment of weighing sacks of sugar until we have determined the weights of

1,000 sacks. With this much data we can decrease the width of the ranges from 1 lb (as in Fig. 2-1) to 0.1 or even 0.01 lb. Larger amounts of data will allow even smaller increments of weight. Figure 2-2, shows the results that we might obtain from the weighing of 1,000 sacks. The mean value of weight is 100 lb and the standard deviation is 2.0 lb.

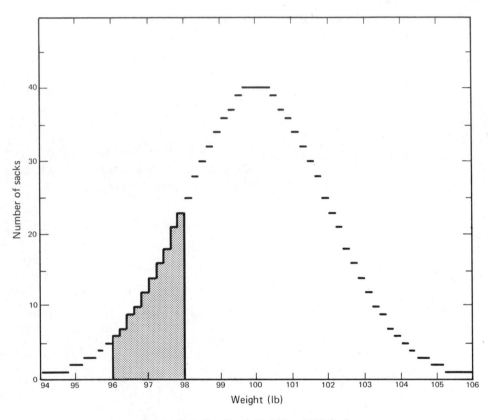

Figure 2-2. Results of Weighing 1000 Sacks

Let us consider the area of the bars in Fig. 2-2. Notice that the height of a bar represents a number of observations. The width of the bar gives the size of the range of weight in which the observations lie. We interpret the area of a bar to be proportional to the number of observations which are in some particular range. For instance, the shaded area in Fig. 2-2 represents the number of sacks that have weight between 96 and 98 lb. If you check the figure you will find that this is 136 sacks. What do you suppose the area of all of the bars represents? Since we have plotted the results of all the observations, the total area must be proportional to the number of sacks that were weighed (1,000 in this case).

Now suppose that you are asked how many of the next 100 sacks weighed will have weights between 99 and 102 lb. What shall we do? First let us assume that the next 100 weights will be scattered or distributed exactly as those that were used in making Fig. 2-2. Since we know that area under the curve represents a number of observations, we can determine the fraction of sacks that weighed between 99 and 102 lb by comparing the area of the bars between these weights with the total area under the curve.

The ratio of the area in question to the total area is 0.533. Thus 53.3 percent of the sacks weighed were within the specified range of weight. If nothing changes, we can state that 53 or 54 of the next 100 sacks will weigh between 99 and 102 lb. Notice that there is no way to determine exactly which sacks will have the specified weight. Statistics is a useful and powerful tool, but we cannot work miracles with it.

We now need to consider the concept of *probability*. In discussing probability, we have one immediate problem. There is some disagreement as to exactly how this term should be defined. Each of the various viewpoints has merits. Rather than become involved in these arguments, we will use a definition that is quite useful and workable, although it may have some theoretical disadvantages.

We will define *probability* as the likelihood that some event will occur. There are a number of ways that we can arrive at a numerical value for a particular probability. One of the most common is to rely upon the results of some experiment. We can use our sugar sack weighing as an example. Consider Fig. 2-2. If we wish to determine the probability that a sack weighs between 98 and 99 lb, we would divide the number of sacks observed to have weights in this range (149 in this case) by the total of observations (1,000). This relative frequency of occurrence is taken as the probability that a sack weighs between 98 and 99 lb.

Perhaps it would be helpful to define the probability a little more formally. The probability that an event will occur (P_i) is defined as

$$P_i = \frac{n_i}{N} \tag{2-8}$$

where

$$n_i = \text{number of times the event is observed}$$
$$N = \text{total number of observations}$$

If we consider Fig. 2-2 again, we see that taking the ratio of areas yields the equivalent of the above definition.

Note that the probability of an event is never negative. The minimum value of P_i is zero; this is the case when the event we are examining never occurs. The maximum value of P_i is one; this is the case in which the event we have chosen always occurs.

Based on Fig. 2-2, what is the probability that a sack weighs between 94 and 106 lb? Since this range includes all of the observations, then

$$P = \frac{n_i}{N} = \frac{1000}{1000} = 1.0$$

We can make the ordinate proportional to probability by simply dividing by the total number of observations. This will make our graph much easier to use.

Did you notice that there appears to be some relationship between weight and number of observations in Fig. 2-2? In fact, we could easily draw a smooth curve passing through the top of each bar. Upon further investigation, we would find that the equation for this curve is

$$f(x) = \frac{1}{\sqrt{2\pi}\sigma} \exp\left(-\frac{1}{2}\left[\frac{x - m}{\sigma}\right]^2\right) \qquad (2\text{-}9)$$

where

$$x = \text{weight of sack}$$
$$m = \text{mean value of weights}$$
$$\sigma = \text{standard deviation, as calculated from Eq. (2-7)}$$

This equation describes the normal or Gaussian distribution. Although it may not be apparent, we have simplified computation of probability by using Eq. (2-9) to approximate our results. If we integrate between the limits $x = \pm\infty$, we find that the value of the result is one. We should expect this since the curve is related to the probability that the weight is within some range. We can calculate the probability that a sack of sugar will weigh between x_1 lb and x_2 lb by integrating $f(x)$ between the values x_1 and x_2.

We must attend to only one other detail. If you try to integrate Eq. (2-9) you will have a great deal of difficulty. It is not feasible to integrate this function every time we need a probability. To avoid this, we can use tabulated values of the results of integration of this function. As σ and m might have any value, it is necessary to make a simple transformation of Eq. (2-9). We define a variable z as

$$z = \frac{x - m}{\sigma} \tag{2-10}$$

When we use this variable to eliminate x from Eq. (2-9) the result is

$$f(z) = \frac{1}{\sqrt{2\pi}} e^{-z^2/2} \tag{2-11}$$

which is the probability density function for a normal distribution with $\sigma = 1$ and $m = 0$. Notice that any normal distribution can be represented in this way by using the transformation $z = (x - m)/\sigma$.

Since we can express all normally distributed events as though $m = 0$ and $\sigma = 1$, it is necessary to set up only one table of areas under the normal curve. There are several ways that this table can be arranged. One procedure would be to integrate Eq. (2-11) between the limits $-\infty$ to various values of z. The resulting table would contain the probability that a result is less than or equal to the selected value of z. A second method (the one I have selected) takes advantage of the symmetry of the normal curve. Equation (2-11) is integrated from zero to various positive values of z. The resulting table gives the probability that a result is between zero and the selected value of z. Since the curve is symmetric, the area from zero to z is the same as the area from $-z$ to zero. This table is presented in the appendix. The following example will demonstrate the use of the table.

The mean weight of 2,000 bearings is 20 oz. The standard deviation is $\sigma = 2$.

(a) How many bearings would you expect to weigh between 16 and 23 oz? We first compute the value of z.

$$z_1 = \frac{16 - 20}{2} = -2$$

$$z_2 = \frac{23 - 20}{2} = 1.5$$

From the probability table, we find that the area corresponding with $z = 2$ is 0.4773. The probability that a bearing weighs between 16 and 20 oz is 0.4773. Similarly, the probability for $z = 1.5$ is 0.4332, so we expect 43.32 per cent of all bearings to weigh between 20 and 23 oz. The probability that a bearing weighs between 16 and 23 oz corresponds to the sum of the two areas. Thus, we expect the fraction of bearings that weigh between 16 and 23 oz to be $0.4773 + 0.4332 = 0.9105$. In a set of 2,000 bearings, we expect 1,821 to be within the specified range of weight.

(b) How many bearings would weigh over 25 oz? Here we are interested in the area between

$$z = \frac{25 - 20}{2} = 2.5 \text{ and } z = \infty.$$

This can be determined by subtracting the area between $z = 0$ and $z = 2.5$ from 0.5000, which is the area between $z = 0$ and $z = \infty$.

$$P = A = 0.5000 - 0.4938 = 0.0062$$

Thus, in 2,000 bearings we expect 12 or 13 to weigh more than 25 oz.

By now you should have developed some understanding of the relationship between the standard deviation and the way indications are distributed about the mean value. To clarify this, let us consider the percentage of a large group of normally distributed items that will be within $\pm\sigma$, $\pm 2\sigma$, and $\pm 3\sigma$ of the mean.

If $X_1 = m - \sigma$ and $X_2 = m + \sigma$,

$$z_1 = \frac{m - \sigma - m}{\sigma} = -1 \quad \text{and} \quad z_2 = \frac{m + \sigma - m}{\sigma} = 1$$

From the table, $A_1 = A_2 = 0.3413$. Thus, we could expect 68.26 percent of the items to be found within $\pm\sigma$ of the mean.

In the case of $X_1 = m - 2\sigma$ and $X_2 = m + 2\sigma$, we find that 95.46 percent of the data will be included. For $X_1 = m - 3\sigma$ and $X_2 = m + 3\sigma$, 99.74 percent of the items are included.

We frequently speak of the *probable error*. This is the interval about the mean in which we expect to find half of the data. In terms of area under the normal curve, we are interested in the symmetric area about the mean value which has a value of 0.5. Since only half of this area is to the right of $z = 0$, we seek the value of z in the table that corresponds to $A = 0.25$. This value is $z = 0.6745$.

Now

$$\pm z = \frac{\text{P.E.}}{\sigma}$$

in this case. Thus,

$$\text{P.E.} = \pm 0.6745\sigma. \tag{2-12}$$

This discussion applies to those cases in which N is large, say 100 or larger. In calibration (that is how this discussion started, remember?), we usually repeat an input only a few times (four or five). Does our previous discussion still apply? If we can assume that our readings are from a normally distributed population, it does.

PROBLEMS

2-3 For the data in problem 2-2, determine the number of balls expected to be greater than 0.503 in. and less than 0.497 in. in diameter. How many balls are expected to have diameters within $\pm.001$ in. of the mean diameter?

2-4 If the students in Prob. 2-1 are a representative sample of the male students on campus, what is the probability that the football coach will select seven men at random who have an average weight of 200 lb or more?

2-5 Within what range will the weights of 60 percent of all male students lie, given the data from Prob. 2-1?

**2-3
Combined
Distributions**

In many cases, we are concerned with characteristics that are functions of normally distributed parameters. It is necessary to determine the standard deviation of the resulting distribution. This is a very complex matter that is difficult to deal with exactly. In most cases, it will be safe to assume that the distribution is normal. Let us see how we can compute the mean value and standard deviation of the distribution.

Suppose we have two variables, x and y, which are normally distributed. The mean values ($\bar{x}$ and $\bar{y}$) and standard deviations (σ_x and σ_y) are known. A third variable, z, is known to be a function of x and y. That is,

$$z = f(x, y)$$

What is the mean value ($\bar{z}$) and standard deviation (σ_z) of the distribution? The mean value is taken to be the value of z when $x = \bar{x}$ and $y = \bar{y}$.

$$\bar{z} = f(\bar{x}, \bar{y})$$

The standard deviation of z can be calculated from the following relationship:

$$\sigma_z = \sqrt{\left(\frac{\partial z}{\partial x}\right)^2 \sigma_x^2 + \left(\frac{\partial z}{\partial y}\right)^2 \sigma_y^2} \qquad (2\text{-}13)$$

The partial derivatives are evaluated at $x = \bar{x}$ and $y = \bar{y}$.

To illustrate these ideas, suppose we are using a manometer to measure pressure. From Chapter 1, we know that

$$p = \gamma h$$

We find that $\gamma = 62.4\ \text{lb/ft}^3$, $\sigma_\gamma = 0.3\ \text{lb/ft}^3$, $h = 1.3\ \text{ft}$ and $\sigma_h = 0.01\ \text{ft}$. What is P and σ_p? From Eq. (2-13)

$$\sigma_p = \sqrt{\left(\frac{\partial p}{\partial \gamma}\right)^2 \sigma_\gamma^2 + \left(\frac{\partial p}{\partial h}\right)^2 \sigma_h^2}$$

Since $(\partial p/\partial \gamma) = h$ and $(\partial p/\partial h) = \gamma$,

$$\sigma_p = \sqrt{(1.3)^2(0.3)^2 + (62.4)^2(0.01)^2}$$
$$\sigma_p = 0.41\ \text{lb/ft}^2$$

The mean value of p is

$$p = \gamma h = 62.4(1.3) = 81.1\ \text{lb/ft}^2$$

PROBLEMS

2-6   The dimensions of a triangle are measured and found to be

$$b = 3.5\ \text{in.}$$
$$h\ (\text{altitude}) = 6\ \text{in.}$$

The standard deviation for both dimensions is 0.03 in. What is the area of the triangle and its standard deviation?

2-7   A tank that is 5 ft long, 3 ft wide and 2 ft high is filled with vegetable oil ($\gamma = 58$ lb/ft³). If $\sigma_L = 0.05$ ft for all dimensions and $\sigma_\gamma = 0.1$ lb/ft³, how many lb of oil are in the tank?

2-4
The Best-Fitting
Straight Line

As we will see in Chapter 3, we often find it convenient to represent the results of a static calibration as a straight line. Due to the scatter that we normally encounter, the calibration points are rarely in a straight line. How do we decide which of the many possible lines best represent the data?

The least-squares method

Assume that we have the set of data which appears in Fig. 2-3. The equation of a straight line which we wish to fit to these data is

$$y = mx + b \qquad (2\text{-}14)$$

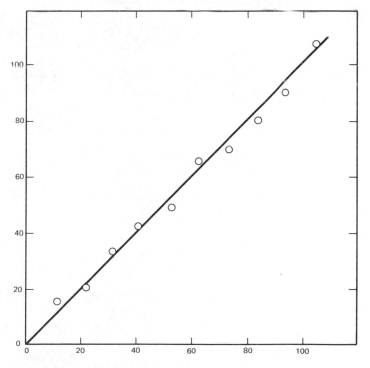

Figure 2-3. A Straight Line Fit to Data

Of the many lines we could draw, which one fits the data best? Before we can answer this satisfactorily, we must agree upon some criterion by which to judge the suitability of any line we might choose. The most generally adopted criterion is based upon minimizing the sum of the squares of the distances from the line to the data points. This may sound formidable, but as you will see, it is not so difficult.

We must now define the distance from the line to each point. The minimum distance is, of course, along a line normal to the line $y = mx + b$ that we are trying to find. In many cases, this is the value we should use. In our case (we are calibrating an instrument) the input is known quite precisely. We can safely assume that any deviation from linearity occurs because of variation of the output quantity. Since we know x much better than y, it is reasonable to choose to minimize the vertical distance from the point to the line. For a particular data point (x_i, y_i) for instance, the distance is given as

$$s_i = y_i - (mx_i + b)$$

The function we wish to minimize is

$$S = \sum_{i=1}^{n} s_i^2 = \sum_{i=1}^{n} [y_i - (mx_i + b)]^2 \qquad (2\text{-}15)$$

where n is the number of points. Since m and b are the variables we wish to determine, the function, S, is minimum when

$$\frac{\partial S}{\partial m} = 0 \quad \text{and} \quad \frac{\partial S}{\partial b} = 0$$

In other words

$$\frac{\partial}{\partial m} \sum_{i=1}^{n} [y_i - (mx_i + b)]^2 = 0 \qquad (2\text{-}16)$$

and

$$\frac{\partial}{\partial b} \sum_{i=1}^{n} [y_i - (mx_i + b)]^2 = 0 \qquad (2\text{-}17)$$

Performing these operations and solving for m and b gives

$$m = \frac{n\Sigma(x_i y_i) - \Sigma x_i \Sigma y_i}{n\Sigma x_i^2 - (\Sigma x_i)^2} \qquad (2\text{-}18)$$

$$b = \frac{(\Sigma y_i)(\Sigma x_i^2) - (\Sigma x_i)\Sigma(x_i y_i)}{n(\Sigma x_i^2) - (\Sigma x_i)^2} \qquad (2\text{-}19)$$

PROBLEM

2-8 For the following sets of numbers, assume that y is a linear function of x and solve for the line that best fits these data using the least squares method. Plot the line and the points to a suitable scale.

(a)

x	0.0	0.1	0.2	0.3	0.4	0.5	0.6	0.7	0.8	0.9	1.0
y	0.01	0.05	0.09	0.13	0.18	0.20	0.25	0.27	0.33	0.38	0.41

(b)

x	0	2	4	6	8	10
y	1.9	29.9	57.7	86	113.6	142.5

x	0	2	4	6	8	10
y	1.8	29.5	58.2	85.3	115.1	142.3

(c)

x	10	12	14	15	18	20	26	30	34	42
y	55	71	85	94	102	118	145	177	203	253

Standard error of estimate and confidence level

We now have a method of deriving the equation of a straight line that best fits a set of data. We are aware that this line does not pass through all (or perhaps any) of the individual points. If we generate another piece of data in some appropriate fashion, the probability is rather high that it will not be on the derived line, either. How can we describe this situation? Is there a way that we can predict some range in which we expect this or any other point to be? The answers to these questions are covered in the following paragraphs.

Suppose the calibration points are distributed in a random fashion about our best-fit curve $y = mx + b$. For any x, this equation gives the mean value of the distribution. Assume also that the distribution (that is, the standard deviation) of the data is not a function of x. In view of these assumptions, we can define a standard deviation for this set of data. To avoid confusion with the standard deviation previously defined, I will call this the *standard error of estimate*, s_e

$$s_e = \sqrt{\sum_{i=1}^{N} \frac{(y - mx_i - b)^2}{n - 2}} \tag{2-20}$$

If we assume that the calibration outputs are normally distributed about the mean, we can use a probability table to estimate the probability that a calibration point will be within a given distance of the calibration curve. For instance, if we define a region bounded by lines $\pm s_e$ from the calibration curve, 68.27 percent of all calibration points will be included. A range of $\pm 0.6745\, s_e$ will include half of the points.

Knowledge of the standard error of estimate allows us to state the probability that a measured point will be within some range of the value predicted by the least squares fit line. Adoption of standard terminology will simplify further discussion of this topic. We usually refer to the probability we select as the *confidence level*. A confidence level of 80 percent simply means that the probability that a measured point will be within a specified distance of the computed line is 0.8 (we are 80 percent sure of being correct). The specified distance is generally called the *confidence interval*. The upper and lower bounds of the confidence interval are called the *confidence limits*. These definitions are illustrated in Fig. 2-4.

Is it obvious that s_e, the confidence level, and the confidence interval are not independent? It should be. If the distribution is known (we have assumed it to be normal) and the standard error of estimate is computed, the selection of a confidence level dictates the value of the confidence interval. The relationship among these quantities is given by Eq. (2-9).

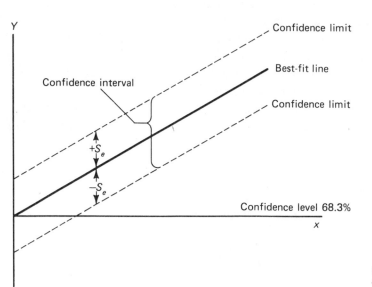

Figure 2-4. Best-Fit Line with Confidence Limits

Finally, we can conclude that a least squares fit line should be presented as shown in Fig. 2-4. The user of this curve will know, for a given value of an independent variable, the mean value of the measured value of the independent variable plus the probability that an individual point will lie in a given range about the mean value. He can also determine, if he knows that the points have been normally distributed about the mean, the confidence interval that will include any fraction of points he desires.

PROBLEM

2-9   For the data given in Prob. 2-8, compute the standard error of estimate, s_e. On the curve, plot the range in which you would expect to find 90 percent of all outputs.

Before proceeding to other topics, let us note that the discussion in the preceding sections has been limited to those cases in which the relationship between two variables is of the form

$$y = mx + b$$

We have further restricted our discussion to those cases in which the data points are normally distributed about our best-fit line and for which the standard error of estimate is independent of x. This restricted case is sufficient for the purposes of this book but we should realize that these concepts can be extended to more complicated situations.

3

The Static
Performance of
Instruments

Suppose an instrument salesman finds his way to your office and is trying to persuade you to buy some of his wares. He begins describing his product in terms such as accuracy, linearity, sensitivity, and resolution. What does he mean? How were these values determined? Is 'infinite resolution' a good or a bad attribute? The answers to these (and other) questions are contained in this chapter.

When you have completed this chapter you should be able to:

1. Describe the procedure known as static calibration. (3-1)
2. Define the term *standard*, list the various classes of standards in order of accuracy, and state the accessibility of each. (3-1)
3. When given the output of an instrument for repeated application of any fixed input, calculate the probability that the next output will be within any given range of values. (3-2)
4. Plot a calibration curve from the results of a static calibration and compute and plot the appropriate confidence limits. (3-4)

5. For each of the following static characteristics, give the definition of the term and, when given appropriate data, calculate its value:
   (a) *bias* (3-3)
   (b) *precision* (3-3)
   (c) *error* (3-5)
   (d) *static sensitivity* (3-5)
   (e) *threshold* (3-5)
   (f) *resolution* (3-5)
   (g) *linearity* (3-5)
   (h) *hysteresis* (3-5)

In Chapter 1 we explored the functions of various components of instruments, and therefore we have some idea of how an instrument operates. Now we will consider how well an instrument performs its various functions. By what measure can the performance of an instrument be judged? How does a "perfect" instrument behave? Ideally, an instrument should present an output that is exactly equivalent to the input imposed upon it. Although no instrument behaves this nicely, this "perfect" instrument will be the ideal by which we measure the performance of real instruments.

Instrument performance is described by means of quantitative qualities that I will call *characteristics*. These characteristics are classified either *static* or *dynamic* depending upon the input imposed when the characteristic is measured. We will consider static characteristics and static performance in this chapter and dynamic performance in Chapters 4 and 5.

**3-1
Static
Calibration**

The process by which we measure the static characteristics of an instrument is *static calibration*. This is accomplished by imposing known, constant values of input upon the instrument and observing the resulting output. This process seems simple enough, but there are some difficulties to consider. These are:

1. The instrument's sensitivity to inputs other than the desired input.
2. The difficulty in obtaining "known constant values" of the input quantity.
3. The failure of most instruments to have the same output for repeated applications of any particular value of input.

Obviously, we must consider and eliminate each of these difficulties in order to obtain a reliable calibration.

Sensitivity to other inputs

Unfortunately, many instruments will present an output for input quantities other than the one in which we are interested. For example, certain devices that are designed to measure pressure are also sensitive to acceleration. We cannot distinguish between an output that is due to a pressure variation and an output that is caused by acceleration. To avoid errors in calibration, it is necessary to control these unwanted inputs so that any change in output of the instrument is due only to changes of the measured input. Usually this can be accomplished by carefully controlling the environment in which the calibration is performed.

In some instances it is necessary to know the static characteristics of an instrument, when the unwanted inputs are varied singly and the primary input is maintained at a constant value. As you can see, it may be necessary to perform several static calibrations of the instrument, one for each possible input. For the remainder of our discussion, let us assume that we need to be concerned only with the primary input and that no unwanted inputs are present.

Standards

How can we obtain known values of input for use in static calibration? To be more specific, suppose you have been assigned the task of calibrating a platform scale. How do you locate weights (masses) to use as input values? It is not too difficult to find constant masses, such as pieces of scrap from a junk yard. Now, how do you decide the numerical value of each of these masses? This problem is a little more difficult, but you can solve it by taking your masses to a second scale, which has been properly calibrated. As the mass of each is measured, you can mark it and you have your own set of calibration masses.

In what way was the second scale calibrated? Do you think that someone somewhere has a master set of masses of known value? If so, who has them and how did he get them? Is there some way you can gain access to them? Can you answer each of these questions? If so, you can skip to the next section. If not, read on and discover the answers.

There is indeed a mass of known value (we will see where in a moment). This mass was defined as a certain quantity of platinum by an international convention. We now have a standard of mass, that is, a quantity of mass that can be used as a reference in determining the values of unknown masses. This defined mass is referred to as the *International Prototype Standard*. From the International Prototype Standard are derived the *National Prototype Standards* maintained by various countries. In the United States, the National Bureau of Standards is charged with maintaining the various standards. Since the national prototype standard is derived by comparison with the international prototype, there is some chance that its value will be somewhat in error. This error is very small, but we must remember that no measured quantity can ever be known exactly.

To protect the National Prototype Mass from deterioration due to abrasion, oxidation, and the like, it is not routinely used. Rather, a third level of standard, the *National Reference Standard Mass* is derived from the National Prototype. In turn, *Working Standards* are developed by comparision with the Reference Standards.

All these standards are used and maintained under carefully controlled conditions. They are not intended to be used under the conditions normally encountered in manufacturing plants and commercial and university laboratories. The standard normally used in these circumstances is the *Interlaboratory Standard*. This standard has a calibration that can be traced to the appropriate standard at the National Bureau of Standards. You should be aware that the value of the Interlaboratory Standard, as well as those of the Prototype, Reference and Working Standards are not known exactly, but are guaranteed to be within some specific range.

Referring to the previous description, we can answer our original set of questions more completely. The National Bureau of Standards maintains a standard mass, which is derived from the mass defined by international agreement. We do not have direct access to this standard, but we can obtain interlaboratory standards, which have calibrations that can be traced through the various levels of standards to the national prototype.

A number of standard quantities have been determined by international agreement. Among these are mass, length, time, a temperature scale, and certain electrical quantities. The previous discussion applies equally to each of them.

Although it is not essential to our studies, we will discuss the numerical values (units) that are assigned to the various standards. You are familiar with the *customary* or *English* system of units that is currently in use in this country. From the state of our technology it is obvious that this is a workable system. There are two disadvantages to this system. First, the variety of units assigned to a particular quantity requires a great number of conversion factors. (Consider length, for instance: inches, feet, yards, miles, rods, fathoms, etc.)

A more important disadvantage is that the United States is virtually the only nation that uses this system. Most other nations use some form of the metric system. In this system, units for the same quantities vary by some power of ten. (For length we have millimeter, centimeter, meter, kilometers, etc.). At the present time, there are some differences among some units of the various national metric systems. In an effort to eliminate these differences and other inconsistencies in the metric system, the *International System* of units (SI) has been proposed. Although this system is not yet completely developed, it is expected that it will be adopted by all nations, including the United States.

Now that we have settled the question of obtaining known inputs for static calibration, let us consider the third problem involved in calibration: the failure of most instruments to have the same output for repeated application of any particular value of input.

PROBLEM

3-1   Find and state the way that standard values for mass, length, time and temperature are defined.

3-2
Response to
a Repeated
Value of Input
Now that we have obtained known constant values of input, let us see what happens if we attempt a static calibration. Referring again to the problem of calibrating a scale, we select a standard weight (say 100 lb), place it on the scale, and record the indication of the scale. As we know that several trials are better than one, we repeat the measurements several times and obtain the following results:

Trial	Output
1	102.3
2	99.3
3	100.9
4	99.1
5	99.9
6	101.4
7	97.5
8	98.2
9	97.5
10	99.5

At this point we probably regret not stopping with a single reading. None of the indications are correct. The average of the readings (99.56) is not the value of the input. What do we do? One solution would be simply to report the data as shown. This is not a solution, but rather a shifting of responsibility. Surely we can do better than that. What information should we report? Certainly the average result, even though we might have obtained a variety of individual indications, is useful.

How should we deal with these individual readings? Suppose it is possible to make only one reading. Is there any way to indicate where this reading might be with respect to the mean? You have probably realized by this point that we are going to have to resort to some statistical description. We will speak only of calibration at present, but remember that these principles also apply to measurement of unknown quantities.

As a first step in considering our problem we must assume that for each possible input (weight) placed on the scale, there are an infinite number of possible outputs. Making a measurement, then, is a matter of obtaining one of the many possible outputs (somewhat like drawing numbered beans from a jar). Unlike drawing beans, we are more likely to obtain certain outputs than others. That is, the outputs are distributed in some nonuniform fashion.

What is the nature of this distribution? We know that the scale has mechanical limitations, which prevent the output from deviating from the input by an amount as large as infinity. Thus, there are finite upper and lower limits that the output cannot exceed. Experience indicates that most of the outputs will be grouped near the mean value. Exact description of the distribution would be quite difficult. A few simplifying assumptions will allow us to use a standard distribution that will adequately approximate the real distribution.

We must first assume that the range of outputs is from negative to positive infinity. This may not seem reasonable, but we will see that the chances of obtaining such extreme outputs are so small that the effect of assuming their possibility can be neglected. A second assumption is that the factors that cause the output to be in error are random and independent of one another. Finally, we assume that outputs in excess of the mean value of output are as likely to occur as those that are less than the mean. As you may have guessed by now, these assumptions allow us to use the normal distribution that was discussed in Chapter 2 to approximate the distribution of outputs from a real instrument.

Let us make sure that assuming the possibility that extreme outputs can occur does not cause any difficulty. Refer to Fig. 3-1. The unshaded area represents the outputs we expect from our instrument. The shaded area represents the additional outputs that will occur if the range of output is allowed to be $\pm\infty$. What is the probability that an output will be in the shaded region? If we let $z = 3.5\sigma$, we find that the shaded area is 0.0004. This means that the probability that we would obtain an output outside the limits of the instrument is 0.0004. This is, of course, small enough that we can neglect the possibility entirely.

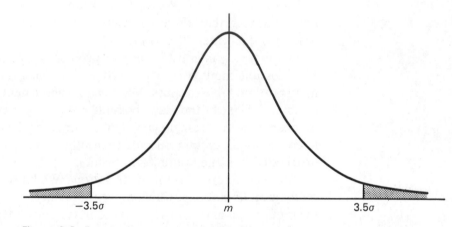

Figure 3-1. Error in Assuming a Normal Distribution

We can now describe the results of our calibration. The average value of the ten readings gives us an approximation of the mean value of the distribution of all outputs. The standard deviation is a measure of the way the outputs are scattered about this mean value. Simply reporting these two values is adequate, but it is more convenient to state the information in a different form. Let us consider the data to illustrate this. The average value of these ten indications is 99.56 lbs. Now compute the standard deviation. Did you get $\sigma = 1.6$? Good. Now, what is the range in which we expect to find 90 percent of all outputs for an input of 100 lb? Recalling the information in Chapter 2, we refer to the probability table in the appendix. There we find that $z = 1.645$ for $A = 0.45$. Since

$$\pm z = \left| \frac{x - m}{\sigma} \right|$$

$$x = m \pm z\sigma$$

$$x = 99.56 \pm 2.63$$

If we know that the above was computed for $P = 0.9$, the above statement can be interpreted as follows: for an input of 100 lb we can expect the output to be between 102.19 lb and 96.93 lb 90 percent of the trials.

PROBLEM

3-2　For the example just discussed, what is the probability that an output will be in the range 99.56 ± 1.00 lb if the input is 100 lb?

3-3　A pressure of 100 psi is applied seven times to a pressure gage. The seven outputs are:
97 psi, 99 psi, 94 psi, 96 psi, 98 psi, 96 psi, & 97 psi.
Specify the range in which you would expect 80 percent of all outputs to appear.

3-4　The probable error of an accelerometer is specified as 2.0 ft/sec². What is the standard deviation? What is the probability that an output is within ± 2 ft/sec² of the mean value of output for a given input?

3-3
Bias and
Precision
It is necessary to make two definitions at this point. First, we will define the difference between the input and the mean value of the output as *bias*. Often the bias is correctable. That is, we can eliminate it by adjusting the instrument. The second definition involves the scatter of the outputs.

Precision is defined as the ability of an instrument to give the same output for repeated applications of a given input. An instrument of high precision gives outputs which are not widely scattered.

It is convenient to be able to assign numerical values to the bias and precision in order to be able to make more meaningful comparisons between instruments. *Bias* is defined as the difference between the mean value of all possible outputs and the input. Thus, bias would be computed as

$$\text{Bias} = \text{Mean Value} - \text{Input}$$

Note that computing bias in this manner yields positive values when average outputs are higher than the input.

How can we describe precision numerically? From the definition of precision we know that it describes the way individual indications are scattered about the mean value. Since precision is a statistical quantity, we can apply our knowledge of statistics to describe it. If we know the distribution, which we have assumed to be normal, and the standard deviation, which can be calculated from calibration data, we can state the probability that any output will be within a specified range. Thus, to specify precision completely, we must specify both a range of values within which we expect the output to lie and the probability that this will occur. The precision of our scale, with a probability of 0.99, is ± 4.1 lb. That is, 99 out of 100 inputs of 100 lb on the scale platform will be within ± 4.1 lb of the mean value of output (which we calculated to be 99.56 lb).

PROBLEMS

3-5   An instrument is said to have a bias of -1.3 units when the input is 40 units. What is the average output when an input of 40 units is applied?

3-6   What are the bias and precision of the instrument described in problem 3-3?

3-7   Would you prefer a voltmeter with a bias of -1.0 volt and a precision of ± 0.1 volt, or a meter with a bias of $+0.1$ volt and a precision of ± 1.0 volt? Why?

3-4
Presentation of
Calibration
Data

Suppose that we continue our calibration by placing weights of varying value on the platform and observing the output corresponding with each. If we apply each weight several times, the same result occurs as with the 100-lb input. That is, for each input, the individual outputs will be normally distributed about some mean value. The results of a calibration of this nature are plotted in Fig. 3-2. In Fig. 3-2(a), all of the data have been plotted. In Fig. 3-2(b), the mean value and the maximum and minimum values of output have been plotted for each input.

The calibration curve

The points plotted in Fig. 3-2 give a picture of the static behavior of the scale, but it is incomplete. Suppose we need the output for some

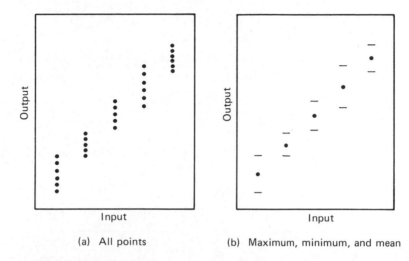

(a) All points   (b) Maximum, minimum, and mean

Figure 3-2. Calibration Data

input other than those appearing in the figure. Can we interpolate between adjacent points? If so, how? In other words, what is the relationship between input and output?

The mean values seem to be distributed about a straight line. Perhaps an assumption that output is proportional to input would be reasonable. There are methods of checking the validity of this assumption for a given set of data, but we will not explore these at this time. Let us simply apply our intuition and the fact that a manufacturer will usually attempt to make the static input-output relationship for his instrument linear, and proceed on that assumption.

We now have the problem that was discussed in Chapter 2. How do we select the straight line that best fits our data? You will recall that we decided to select the line that minimizes the sum of the squares of the distances from the individual points to the line. Since we are calibrating the instrument, we can assume that the uncertainty lies in the output rather than in the input. In this case, we can use the procedure outlined in Chapter 2.

That is, we assume the output, y, is related to the input, x, by

$$y = mx + b$$

where

$$m = \frac{n\Sigma(x_i y_i) - \Sigma x_i \Sigma y_i}{n\Sigma x_i^2 - (\Sigma x_i)^2}$$

$$b = \frac{(\Sigma y_i)(\Sigma x_i^2) - (\Sigma x_i)(\Sigma x_i y_i)}{n(\Sigma x_i^2) - (\Sigma x_i)^2}$$

and x_i is the value of the input for the ith trial and y_i is the corresponding output.

PROBLEM

3-8 During the calibration of a platform scale the following results were obtained:

Input	Output	Input	Output
10	12.3	10	10.7
20	19.3	20	15.7
30	30.9	30	31.1
40	39.1	40	43.6
50	49.9	50	50.7
60	61.4	60	60.4
70	67.5	70	72.7
80	78.2	80	79.7
90	87.5	90	89.3
100	99.5	100	101.5

Determine the values of m and b that give the straight line that best fits these data. Plot this curve.

Confidence limits

How do we indicate that our calibration curve is not exact, but rather the straight line that best fits our data? An obvious solution is to plot all of the data with the best-fit line. This is helpful and does indicate to the engineer or student that he should expect some scattering of output values from the instrument. However, it is always useful to have some quantitative measure of this scatter. Can we do so in this case?

Let us assume that the calibration points are normally distributed about our best-fit calibration line, $y = mx + b$, and that the distribution is not a function of input, x. For any x, the equation of the calibration line gives the mean value of the distribution at that input. In view of these assumptions, we can compute the standard error of estimate (s_e) for the population of calibration points. This quantity was defined in Chapter 2 by Eq. (2-20). Remember that the standard error of estimate has the same properties as the standard deviation.

Suppose that we draw lines parallel to the calibration line and $\pm s_e$ units from it as in Fig. 3-3. What can we say about the region bounded

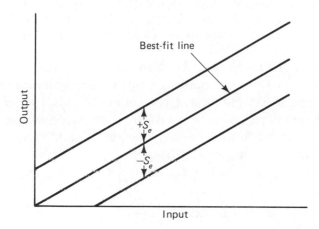

Figure 3-3. Limits of Scatter

by these two lines? From our discussion in Chapter 2, we know that for a particular input to the instrument, the probability is 0.683 that the output will be within the indicated range. Or, taking a slightly different point of view, 68.3 percent of all outputs from the instrument will be within the range $\pm s_e$. With the aid of a probability table we can select the range corresponding to any level of probability we desire. Thus, if we wish to specify the range in which 90 percent of the outputs will lie, we would construct lines $\pm 1.645\, s_e$ from the calibration curve. (Remember that the preceding statement is equivalent to stating that the probability that any given output will be in this range is 0.9.)

Do you recall the terms confidence level, confidence interval, and confidence limits, which we defined in Chapter 2? If not, review them because they apply to our calibration curve.

Before proceeding, let us summarize. After making the assumption that the static input-output relationship is linear, we can determine the equation of the straight line that best fits the calibration data. Our second assumption—that the data are normally distributed about the calibration line—allows us to compute the standard error of estimate. From this we can compute the confidence interval for a given confidence level. The resulting calibration curve shows the mean value of output and the range within which a given fraction of all output values will lie for a given input.

PROBLEM

3-9   For the data given in Prob. 3-8, compute the standard error of estimate, s_e. On the calibration curve, required in Prob. 3-8, plot the range in which you would expect to find 90 percent of all outputs.

The preceding discussion describes how we could predict the output for a given input to an instrument. Consider for a moment how we use an instrument. This is different, is it not? In this case, we know the output and are interested in the input. How can we determine the input? The line that we fit to the data by the least squares method can be used to determine the mean value of x (the input) given a value y (the output) by solving the equation for this line ($y = mx + b$) for x.

$$x = \frac{y - b}{m} \tag{3-1}$$

Can we stop at this point? It certainly would be convenient if we could. However, if you recall that the curve does not pass through all the calibration points, you will agree that we must consider the way all possible inputs corresponding to the measured output y, are distributed about the mean value, x.

To do this, we must establish the confidence limits on the calibration curve when it is used to determine the input corresponding to a given output.

Unfortunately, we have not developed all the concepts of statistics that are required to establish the confidence limits exactly. We can, however, utilize an approximate method, provided that the calibration curve is based on at least 20 points. The equation for the upper and lower confidence limits is

$$CL = x \pm \frac{zs_e}{m} \sqrt{1 + \frac{1}{n} + \frac{(x - \bar{x})^2}{\sum\limits_{i=1}^{n} (x_i - \bar{x})^2}} \tag{3-2}$$

where

$x =$ input corresponding to a particular output, y, as calculated from Eq. (3-1)

$z =$ parameter from probability table corresponding to the confidence level selected

$s_e =$ standard error of estimate from Eq. (2-20)

$m =$ slope of calibration line

Equation (3-2) shows that the confidence limits are not parallel with the calibration line. The confidence interval is narrowest where $x = \bar{x}$ and is broadest at the ends of the line. Since z depends upon the confidence level that is required, the confidence interval also depends upon the confidence level.

PROBLEM

3-10 For the data given in Prob. 3-8, compute the range in which 90 percent of all inputs will lie, assuming that the output quantities are known. Plot this range on the calibration curve you constructed in Probs. 3-8 and 3-9.

We now have two ways of presenting calibration data. We can present the best-fitting straight line and the range in which we expect some fraction of all outputs to lie for a given input, or we can present this same best-fitting straight line and the range in which we expect some fraction of all inputs to lie for a given output. If you are awake and alert at this point, you should be wondering which presentation is "better." This depends on what you are trying to do.

Presenting the range in which the outputs are expected to lie gives a good measure of the precision of the instrument. This kind of information is most useful when you are selecting an instrument for a particular application. Virtually all calibration curves are presented in this way.

The second method of presenting calibration data is most useful when you are interested in predicting the input given the output of an instrument. For instance, if you have such a calibration curve for a pressure gage and obtain an output of 75 psi, you can use the curve to make a statement such as this: the input pressure is 75.2 ± 0.5 psi with a probability of 90 percent.

3-5
Static
Characteristics

A number of measures of performance can be defined and evaluated using the results of static calibration. These will be referred to as *static chatacteristics*. These characteristics describe the performance of an instrument when it is subjected to a constant input. Two of these characteristics, *bias* and *precision*, have been defined. These, you will recall, were defined as the result of the performance of the instrument when a particular input was repeated several times. Most of the definitions that follow are defined using the calibration curve as reference.

Error

By the time you have reached this section, you should be convinced that the output of an instrument is rarely exactly correct. Although the deviation from the correct value is usually small, we are aware that it exists. How shall we specify this discrepancy? *Accuracy* is a term that is sometimes used. It is defined as the ability of an instrument to correctly indicate the value of its input. This is a rather vague term and is usually used qualitatively in making comparisons. Thus, you might say, "Instrument A is more accurate than Instrument B."

Rather than assigning a numerical value to accuracy, we generally elect to define *error* and specify it numerically. Usually *error* is defined as the maximum difference between an output and the output predicted by the calibration curve for the applied input. What is the maximum error? In assuming that the outputs are normally distributed, we assume that the output could be $\pm\infty$. Should we state that the error is infinite? It is highly unlikely that errors of that magnitude will occur, so perhaps we should select a value for maximum error that includes some fraction of all the outputs. That is, if we select a confidence level the maximum error is given by the confidence limits.

Suppose that we have calibrated a pressure-measuring device and found that the standard error of estimate is 0.5 psi. What is the maximum error? To answer this we must select a confidence level. In this case, let us be content to be correct half the time. From a probability table for $p = 0.5$, z equals ± 0.6745. Our confidence interval is $\pm 0.5\,(0.6745) = \pm 0.337$ psi. The maximum error is also ± 0.337 psi.

Often the error is expressed as a percentage of the maximum input that can be imposed upon the instrument. If the pressure gage is designed to operate with input between zero and 50 psi, the error is $\pm 0.337/50 = \pm 0.0065$ or ± 0.65 percent. The statement of error is not complete unless the confidence level is also given.

PROBLEM

3-11   From a static calibration of a thermometer, it has been determined
       that the standard error of estimate is 0.25 degrees. Specify the error.

Thus far we have discussed error for those instruments for which
we can assume that the confidence limits are parallel with the calibration line. This implies that the scatter in output is about the same for
all values of input. This is true for many instruments, but what is done
if, for instance, the scatter increases with increasing input? We will not
present the statistics here, but it should be apparent that for a constant
confidence level, the confidence interval should increase with increasing input. This situation is shown in Fig. 3-4(b). Although the confidence level does not change, confidence limits are a function of input.

How is error specified in this case? We usually assume that the confidence limits are linear, although they diverge from the calibration
line. With this assumption, the confidence interval and therefore the
maximum error are proportional to input. This could be indicated
by specifying the error in terms of the instrument indication; *error:*
$\pm x$ percent of indication.

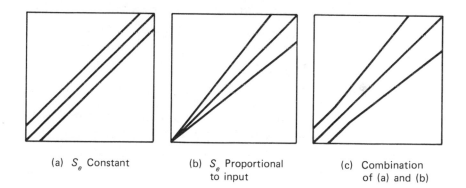

(a) S_e Constant   (b) S_e Proportional   (c) Combination
                                  to input   of (a) and (b)

Figure 3-4. Confidence Interval for Constant Confidence Level

Frequently, the error will be constant over part of the range of the
instrument and proportional to input over the remainder of the range
as shown in Fig. 3-4(c). This is usually indicated by a statement such
as: error equals $\pm x$ units or $\pm y$ percent of indication, whichever is
larger.

PROBLEMS

3-12 The equation of the straight line that best fits a set of voltmeter calibration data is $y = 0.01 + 0.98x$ volts. Show the maximum error and the calibration line for each of the following cases. The maximum input voltage is 250 V.

(a) s_e is constant and is ± 0.1 V.

(b) s_e is ± 0.5 percent of input

(c) s_e is ± 0.5 volts or ± 0.5 percent of input, whichever is greatest.

3-13 What is the maximum error at 120 V for each part of Prob. 3-12?

Static sensitivity

Static sensitivity is also determined from the results of static calibration. This static characteristic states the magnitude of the change in output for a unit change in input.

$$\text{Static sensitivity} = \frac{\text{change of output}}{\text{change of input}}$$

At first, this definition seems rather pointless. After all, the liquid-in-glass thermometer is calibrated to read in degrees and one would expect the static sensitivity to be $1 \dfrac{\text{degree}}{\text{degree}}$. This, though perhaps technically correct, does not convery much information, since every properly constructed thermometer should have a sensitivity of $1 \dfrac{\text{degree}}{\text{degree}}$. If we consider the liquid-in-glass thermometer more carefully, we see that the output is really the displacement of the liquid column. Thus, if a change in input of one degree causes the length of liquid column to change 0.2 in., the static sensitivity is $0.2 \dfrac{\text{inches}}{\text{degree}}$. Defined in this manner, the static sensitivity allows us to compare the sensitivity of various thermometers to changes of temperature. For instance, the static sensitivity of the thermometer in Fig. 3-5(a) is 0.1 in. per degree, and it is one in. per degree 3-5(b) for the thermometer in Fig. 3-5(b).

It should be noted that the static sensitivity is the slope of the calibration curve, *if the ordinate is expressed in actual output units.* If the curve is linear, the static sensitivity is constant. In the case of

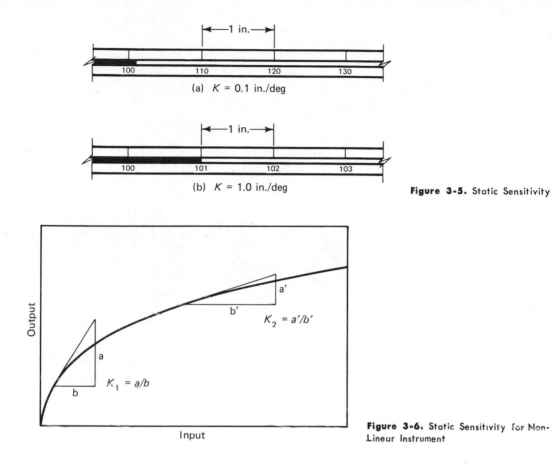

(a) K = 0.1 in./deg

(b) K = 1.0 in./deg

Figure 3-5. Static Sensitivity

Figure 3-6. Static Sensitivity for Non-Linear Instrument

nonlinear instruments, the static sensitivity is not constant and must be specified in terms of the input value. This is illustrated in Fig. 3-6.

PROBLEMS

3-14   The equation of the calibration curve for the pressure gage shown in Fig. P3-14 is

$$y = -1.0 \pm 1.01x$$

The units of x and y are psi. What is the static sensitivity of this instrument?

3-15   What is the static sensitivity of the spring scale shown in Fig. P3-15? Assume that the weight is correctly indicated on the face of the scale. That is, when the applied weight is changed by one lb, the indication also changes by one lb.

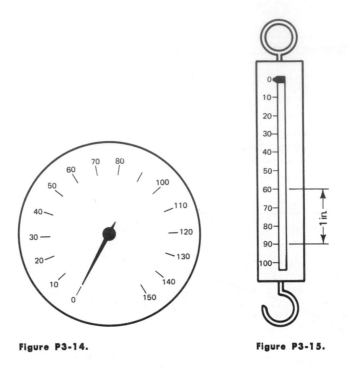

Figure P3-14.

Figure P3-15.

3-16 The calibration curve for an instrument is

$$y = x - 0.001x^2$$

What is the static sensitivity of the instrument at $x = 50$ units? $x = 100$ units?

Threshold and resolution

Two related terms are *threshold* and *resolution*. If we consider a platform scale with range of 0-1,000 lb, we see that it will not, for instance, weigh a fly. In other words, there is some minimum input for which there will be an output. Below this minimum input the scale will indicate zero. We call this minimum value of input for which there will be an output the *threshold* of the instrument. This may be expressed in input units or as a fraction of full scale.

Resolution is the smallest change of input for which there will be a change of output. In other words, if our scale has an input of 100 lb it might not respond to an increase of 0.01 lb, but would perhaps require an increase of at least 0.1 lb before the output would change.

Resolution may also be expressed in input units or as a fraction of full scale.

Threshold and resolution are not zero because of certain details of construction of the instrument. For instance, friction between moving parts and "play" or looseness in joints (more correctly termed *backlash*) contribute to loss of resolution.

Linearity

Although manufacturers attempt to design their instruments so that the output is a linear function of the input, small deviations occur. The term *linearity* describes the maximum deviation of the output of the instrument from a best-fitting straight line through the calibration points. In speaking of the output in this context, we are referring to the *mean value* of output for each input. Perhaps a simple example will clarify this.

Let us assume that we have calibrated an instrument and fit a straight line, $y = x$ to the data. (y is the output and x is the input). Suppose that the actual relationship is

$$y' = x + 0.01 \sin x$$

The distance between the actual calibration line and the least squares fit line for any value of x is

$$d = y' - y$$

For our example

$$d = x + 0.01 \sin x - x$$
$$= 0.01 \sin x$$

This distance is maximum when $x = \pi/2, 3\pi/2, 5\pi/2 \cdots$ The value of this maximum distance is 0.01 units.

Although this deviation from the straight line is an expression of *nonlinearity*, the term *linearity* is also used. The statement "linearity = 1 percent" is equivalent to "nonlinearity = 1 percent."

It is frequently not convenient to determine the equation of the actual relationship between input and output. How can we determine the nonlinearity? Let us consider our calibration curve again. For each value of input we have several output values, from which we can compute the mean value of output for each input. We can reasonably assume that each of these average output values is a point on the actual curve relating the input to the output. In this case, the non-linearity is the maximum vertical distance from the calibration line to one of the mean values of output, as determined by static calibration.

PROBLEM

3-17   Use the data and results of Prob. 3-8 to compute the linearity of the platform scale. Specify the linearity as a percentage of the full scale input.

To this point, we have discussed the case in which the magnitude of the deviation of the computed calibration line from the actual curve is approximately the same for all input values [see Fig. 3-7(a)]. What if the actual and computed calibration curves appear as in Fig. 3-7(b)? Here, the deviation seems to be a function of the input, increasing as the input increases. Stating the linearity in terms of the maximum deviation would not give an accurate picture of the static performance of the instrument. Consider Fig. 3-8. Two possible ways increasing nonlinearity might be expressed are shown. In Fig. 3-8(a), the non-linearity approaches zero as the input decreases. The outer lines form an envelope that contains both curves. These lines represent the limits of linearity for all values of input. In this case, the linearity is stated as "± a percent of input."

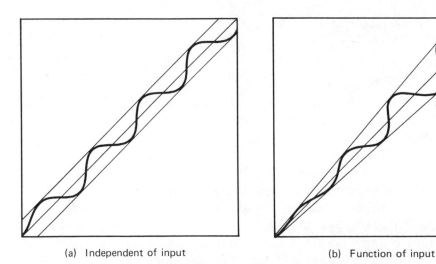

(a) Independent of input   (b) Function of input

Figure 3-7. Examples of Non-Linearity

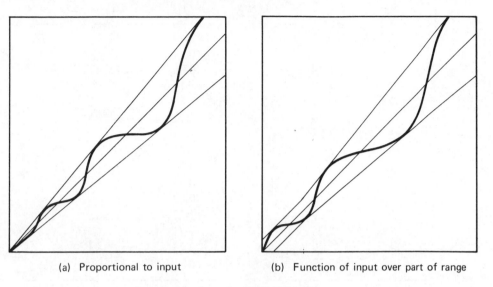

(a) Proportional to input   (b) Function of input over part of range

Figure 3-8. Examples of Non-Linearity

Fig. 3-8(b) shows a case in which the linearity does not approach zero with decreasing input, but approaches some finite limit. Again, we can draw lines that represent the limits of linearity as shown by outer lines in the figure. The linearity can be stated as $\pm a$ units or b percent of input, whichever is greater.

PROBLEM

3-18   The line that best fits the following calibration data is $y - 1.99x + 0.13$. Specify the linearity.

x	y	x	y	x	y
0	-1	0	0	0	-2
20	42	20	39	20	40
40	81	40	80	40	82
60	119	60	121	60	119
80	159	80	161	80	160
100	198	100	198	100	199

3-19   The line that best fits the following calibration data is $y = 0.25x - 0.01$. Specify the linearity.

x	y	x	y	x	y
1	0.13	15	3.72	11	2.85
3	0.82	1	0.14	13	3.29
5	1.24	3	0.86	15	3.74
7	1.79	5	1.13		
9	2.29	7	1.83		
11	2.80	9	2.12		
13	3.25	11	2.79		
15	3.80	13	3.21		
1	0.26	15	3.75		
3	0.81	1	0.29		
5	1.16	3	0.82		
7	1.69	5	1.27		
9	2.28	7	1.71		
11	2.79	9	2.20		
13	3.39				

3-20 The line that best fits the following calibration data is $y = x - .05$. Specify the linearity.

x	y	x	y
0	2	500	497
500	507	400	412
300	299	600	606
100	92	200	207
700	705	200	207
300	291	600	596
600	599	0	−2
200	194	700	706

x	y	x	y
800	799	500	502
100	94	100	102
400	400	700	703
300	308	900	893
900	897	400	390
800	798	0	1
		800	802
		900	902

Hysteresis

Imagine calibrating a pressure gage in the following fashion. First we increase the applied pressure in equal increments until the maximum value of input is attained. The output is recorded for each input value. We then unload the gage by reducing the applied pressure in equal increments, again recording the output for each input value. Rather than combining the data to make a calibration curve, let us plot the points for increasing pressure separately from those for decreasing pressure. You will note that the magnitude of the output for a given input depends upon the direction of the change of input as in Fig. 3-9. This dependence upon previous inputs is called *hysteresis*. We can specify hysteresis as the maximum difference in output that occurs when the input value is first approached from above and then from

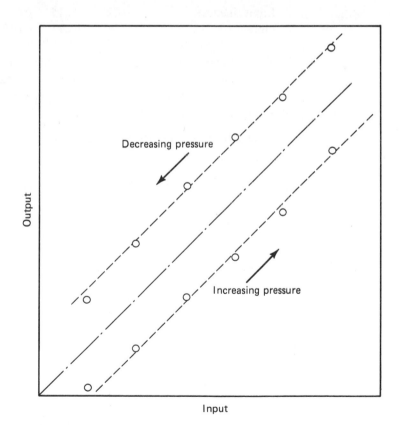

Figure 3-9. Hysteresis

below. The maximum difference is frequently specified as a precentage of full scale.

PROBLEMS

3-21   A dial indicator is calibrated by placing gage blocks of various heights under the spindle and noting the indication of the instrument. The following data are obtained in order to measure the hysteresis. Use these data and calculate the hysteresis.

Increasing heights

Input (inches)	0.001	0.005	0.010	0.015	0.020
Output (inches)	−0.0006	0.0029	0.0079	0.0131	0.0176

Decreasing heights

Input (inches)	0.020	0.015	0.010	0.005	0.001
Output (inches)	0.0182	0.0158	0.0109	0.0062	0.0021

3-22   The following data result from the calibration of a pressure transducer. Notice that the input quantity is pressure expressed in psi and the output is electrical. Use the given data to

1. Calculate precision at an input of 50 psi.
2. Determine the best-fitting straight line through the data.
3. Plot the calibration curve with appropriate confidence limits.
4. Calculate the static sensitivity.
5. Specify the error.
6. Specify the linearity.
7. Specify the hysteresis.
8. Replot the calibration curve with the confidence limits that are appropriate if the curve is to be used to determine the input from the output of the instrument.

The following data were collected in random order. They have been rearranged for your convenience.

Input (psi)	Output (volts)	Input (psi)	Output (volts)
5	0.21	5	−0.15
25	0.47	25	0.38
50	0.55	50	0.58
75	1.33	75	1.61
100	1.84	100	1.92
125	2.28	125	2.62
150	3.07	150	2.99
175	3.71	175	3.24
200	4.05	200	4.21
5	0.24	5	0.12
25	0.47	25	0.73
50	0.54	50	1.29
75	1.75	75	1.60
100	2.09	100	2.04
125	2.55	125	2.39
150	3.11	150	3.15
175	3.21	175	3.56
200	3.85	200	4.25

The following data were taken in the order shown:

Input (psi)	Output (volts)	Input (psi)	Output (volts)
5	−0.07	25	0.41
25	0.74	50	0.75
50	0.78	75	1.66
75	1.43	100	1.94
100	1.73	125	2.39
125	2.58	150	2.82
150	2.80	175	3.58
175	3.39	200	4.29
200	3.59	175	3.43
175	3.30	150	3.13
150	3.23	125	2.67
125	2.33	100	1.78
100	2.08	75	1.44
75	1.33	50	0.93
50	0.99	25	0.44
25	0.67	5	0.10
5	−0.08		

3-6
Examples of
Specifications
of Static
Characteristics
Let us consider the static characteristics of some of the instruments that were introduced in Chapter 1. The characteristics that we are describing are typical but do not describe specific products of any manufacturer. Our purpose is to see how the principles developed in this chapter are applied to real instruments.

Strain gages

Which of the static characteristics discussed apply to strain gages? If we consult a manufacturer's catalog, we will find the following information concerning the gages: the gage factor and its tolerance, the gage resistance and its tolerance, information concerning the maximum elongation of the gage, and information concerning the way the gage factor is affected by temperature change. These are *nominal* or *design* values of the parameters. Because of variations in materials and manufacturing, the values of these characteristics will be somewhat different from the nominal values.

Assume now that we have purchased a package of strain gages. In the package we find a piece of paper with the following information:

Gage Factor at 75°F	2.03 ± 1 percent
Resistance in ohms	120 ± 0.5%
Strain limits	±3%

In addition you may find a curve similar to that shown in Fig. 3-10.

What is the relationship between these terms and the static characteristics we have been talking about in the first part of the chapter?

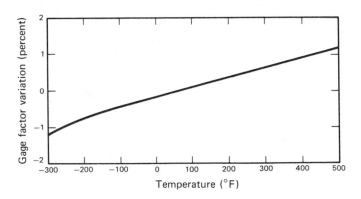

Figure 3-10. Effect of Temperature on Gage Factor

You will remember (from Chapter 1) that strain is related to change in gage resistance by the gage factor;

$$\frac{\Delta R}{R} = \epsilon \, (\text{G.F.}) \tag{3-3}$$

Now, imagine that we have calibrated a strain gage by applying known values of strain and observing ΔR. If the calibration curve were plotted, it would appear as shown in Fig. 3-11. We know that the slope of this line is the static sensitivity. From Eq. (3-3) we see that the gage factor is also the slope of the calibration line and can be considered as the static sensitivity of the gage. The tolerance placed on the gage factor indicates its variability. We can usually assume that a tolerance of this nature is equivalent to three standard deviations (3σ). We can

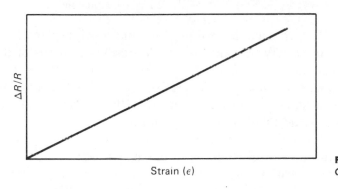

Strain (ϵ)

Figure 3-11. Strain Gage Calibration Curve

interpret the gage factor specification as follows: the gage factor is between 2.01 and 2.05 for 99 percent of the gages offered for sale.

We can make a similar statement concerning the specification of R. That is, the gage resistance is between 119.4 and 120.6 ohms for 99 percent of the gages offered for sale.

The strain limits tell us the range of the gage. We can sense strains of ± 0.03 in. per inch without damaging this particular gage. The curve shows how the gage responds to changes in temperature. Notice that the gage factor is determined at 75° F.

Unbonded strain gages are not usually used to measure displacements, but are usually found as parts of other instruments. We will discuss these a little later.

Accelerometers

The following are some of the static characteristics that might be specified for an accelerometer:

Range	±20 g
Static overload	±200 g
Linearity and Hysteresis	$\pm0.75\%$ FS
Resolution	Infinite
Cross axis sensitivity	0.02 g/g
Operating temperature	-60 to $200°$F
FS output	±4 mV/V open circuit
Thermal zero shift	Less than 0.01% FS/$°$F
Thermal sensitivity shift	Less than 0.01%/$°$F
Bridge resistance	350 Ω
Excitation	7.5 VDC or AC (rms) through carrier frequencies

The first two characteristics do not require much explanation. It is customary to specify the range of accelerometers in terms of the acceleration of gravity. Thus, the range of inputs to which this instrument is sensitive is ±20 times the accleration of gravity. Referring to the statement of static overload, we find that the accelerometer can be subjected to steady state accelerations up to ±200 times the acceleration of gravity without being damaged. This does not imply, however, that we can measure acceleration of this magnitude.

In this specification, the magnitudes of nonlinearity and hysteresis are stated together. The maximum deviation is ±0.0075 times the full scale output. We cannot determine what portion is due to nonlinearity and what portion is due to hysteresis. Since we are usually interested only in the total deviation, this statement is quite adequate.

Although the specification states that the resolution is infinite, we should not be alarmed. This statement means that the resolution is infinitesimal—the output will change for any change of input.

An accelerometer is generally designed to sense acceleration along a single axis. If the direction of this axis of maximum sensitivity is not obvious from the design of the accelerometer, it is marked on the case of the instrument as shown in Fig. 3-12. The cross-axis sensitivity states the sensitivity of the instrument to accelerations normal to this direction. From the specification, an accleration of 1 g normal to the

Figure 3-12. Method of Indicating Direction of Maximum Sensitivity of an Accelerometer. *(Courtesy Kistler Instrument Company)*

axis of maximum sensitivity will result in an output equivalent to 0.02 g directed along the axis of maximum sensitivity.

The statement of FS (full-scale) output will give us the static sensitivity of the instrument. In order to understand this, we need to determine how the input (acceleration) is converted into an electrical signal. If we read on through the specifications, we find a statement concerning bridge resistance. From this, we assume that we are dealing with an instrument that contains some type of four-arm strain gage bridge. (If we had the complete specification, we would not have to make this assumption. The manufacturer would make it clear to us.) If you refer to Eq. (1-6), you will find that the output from a strain gage bridge depends upon the excitation voltage as well as the strain. Now the specification makes sense. When the full-scale acceleration (20 g) is applied, the output is 4 millivolts for each volt of excitation. If the recommended excitation (input) voltage is used, the full-scale output is

$$\pm 4(7.5) = 30 \text{ mV}$$

By definition, the static sensitivity is the change in output per unit change of input. Thus,

$$K = \frac{30 \text{ mV}}{20 \text{ g}} = 1.5 \text{ mV/g}$$

if the input is 7.5 V.

In most cases, we prefer that an instrument have zero output when no input is applied. This is not difficult to accomplish with a resistance bridge, such as that contained in our accelerometer. We find, however, that the electrical resistance of the strain gage is a function of temperature. A temperature change that occurs after zero has been established will often cause the instrument to have an output when no input is applied. We refer to this as *zero shift*. This shift is a vertical translation of the calibration curve as shown in Fig. 3-13. For the instrument

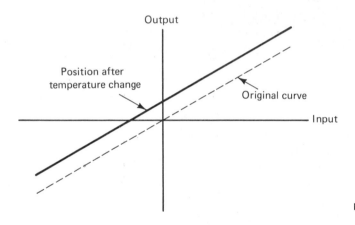

Output

Position after
temperature change

Original curve

Input

Figure 3-13. The Effect of Zero Shift

we are considering, the thermal zero shift is less than 0.01 percent of full scale output for each degree of temperature change. Thus, if zero is established at 70° F and the temperature later increases to 100° F, the instrument will have an output of

$$0.0001(4)(30) = \pm 0.012 \text{ mV/V}$$

when no load is applied. One question remains. In which direction does the zero shift: upward for an increase of temperature and downward for a decrease? The manufacturer does not stipulate this. If the instrument is available, we can determine the direction of this shift by means of a simple experiment. In our present situation, all we can do is state that a temperature change will move the zero no more than $\pm x$ units.

Temperature also affects the static sensitivity of the instrument. This occurs because the gage factor of the strain gage changes with temperature. The result is a change of slope of the calibration curve. According to our specification, this change is less than 0.01 percent per degree F. We can express this as a change in static sensitivity as follows: from the specification of full scale output, we can calculate the nominal value of the static sensitivity. We did this earlier and found it to be 1.5 mV/g with an excitation of 7.5 V. Then the change of static senstivity for, say, a 30-degree change in temperature is

$$\Delta K = 0.0001 \ (1.5)(30) = .0045 \ \text{mV/g}$$

Two pieces of information are missing. We do not know whether the static sensitivity increases or decreases with increasing temperature. We also have no information concerning the temperature upon which the specification of thermal sensitivity is based. Although we cannot be certain, we can probably safely assume that this temperature is 70–75°F.

Figure 3-14 shows the combined effect of zero shift (thermal zero coefficient) and change of static sensitivity (thermal sensitivity coefficient) for a particular change in temperature. Since we do not know what sign to attach to either of these coefficients, we must be content

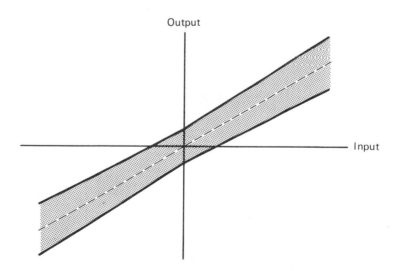

Figure 3-14. Combined Effects of Zero Shift and Change of Static Sensitivity

to state that the calibration curve will be within the shaded area for the stated temperature change.

Other instruments which utilize strain gage bridges in some form will have specifications that are similar to those we have just discussed. We have not included all of the specifications. We have yet to discuss the characteristics that describe dynamic behavior.

PROBLEM

3-23   The following are specifications for a load cell. Explain each of the specifications, using the terms defined in this chapter.

SPECIFICATIONS FOR A LOAD CELL

Range	$\pm 5{,}000$ lb
F.S. Output	2 mV/V
Accuracy	$\pm 0.5\%$ of 2 mV/V
Linearity	Better than 0.3% FS
Hysteresis	$\pm 0.20\%$ FS
Repeatability	$\pm 0.20\%$ FS
Thermal Zero Coefficient	0.0025% FS/°F
Thermal Sensitivity Coefficient	0.003% of Load/°F
Overload Capacity	250%
Bridge Resistance	350 Ω
Maximum Excitation	20 Volts AC-DC
Recommended Excitation	12 Volts AC-DC

We will not attempt to discuss the specification of static characteristics for all of the instruments we discussed in Chapter 1. Rather, you should obtain some typical specifications and study them. I believe they will be intelligible to you by this time.

Before closing this chapter we want to take a look at two instruments that have nonlinear calibration curves. These are the thermocouple and the orifice meter. Figure 3-15 is the calibration curve for an iron-constantan thermocouple. Note that it is definitely nonlinear. The static sensitivity is not constant; it is a function of input temperature.

It is obvious from Eq. (1-12) that the calibration curve for the orifice meter is not linear. A study of the accuracy of this device is also of interest. From Eq. (2-12), which is presented again for your convenience, we see that the error in measurement of flow (Q) is a function of many variables.

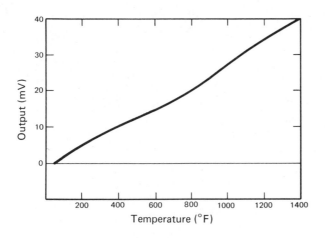

Figure 3-15. Iron-Constantan Thermo-couple Calibration Curve

$$Q = \frac{CA_2}{\sqrt{1 - (A_2/A_1)^2}}\sqrt{\frac{2g}{\gamma}(p_1 - p_2)}$$

Some of the variables that contribute to the error and their own errors are:

Orifice diameter:	± 0.5 percent
Pipe diameter:	± 0.5 percent
Discharge Coefficient	± 0.5 percent
Pressure	± 1.0 percent
Specific Gravity	± 1.0 percent

There are a number of other factors that we might also consider. Rather than confuse ourselves with additional detail, we will proceed on the assumption that the above are the only factors we need to consider.

In Chapter 2 we discussed a method of describing error when a number of variables are involved. From Eq. (2-13),

$$\sigma_Q = \left[\left(\frac{\partial Q}{\partial A_1}\right)^2 \sigma_{A_1}^2 + \left(\frac{\partial Q}{\partial A_2}\right)^2 \sigma_{A_2}^2 + \left(\frac{\partial Q}{\partial C_D}\right)^2 \sigma_{C_D}^2 + \left(\frac{\partial Q}{\partial \gamma}\right)^2 \sigma_{\gamma}^2 \right.$$
$$\left. + \left(\frac{\partial Q}{\partial P_1}\right)^2 \sigma_{P_1}^2 + \left(\frac{\partial Q}{\partial P_2}\right)^2 \sigma_{P_2}^2 \right]^{1/2} \tag{3-4}$$

We can probably assume that the errors given above represent three standard deviations. Although the prospect of differentiating Eq. (2-12) and solving Eq. (3-4) is not pleasant, this procedure gives the best estimate of the error in the measured volumetric flow rate.

PROBLEM

3-24   A sharp edged orifice is used to measure the volumetric flow rate of water. If the errors cited before Eq. (3-4) apply to this meter, calculate the volumetric flow rate and its error.

$$d_1 = 3.068 \text{ in.}$$
$$d_2 = 2.280 \text{ in.}$$
$$\gamma = 62.4 \text{ lb/ft}^3$$
$$P_1 = 15.1 \text{ psi}$$
$$P_2 = 14.7 \text{ psi}$$
$$C = 0.85$$

3-7 Summary

We have discussed a number of measures of performance of an instrument that can be determined by static calibration. We have referred to these measures as static characteristics.

Perhaps the most important concept is that instruments are incapable of giving the same output for repeated applications of the same input. This fact forces us to consider the output of an instrument as a statistical quantity. We must remember that no statement concerning the precision of the instrument is complete unless the confidence level is stated.

The statistical nature of calibration data must be considered when presenting that data. Calibration curves must be prepared to reflect this. We examined a procedure for doing so, based upon the assumption of a linear relationship between input and output. A number of characteristics, based in part upon the assumption of a linear relationship, were defined.

We have studied a few typical examples of the manner in which these static characteristics are specified by manufacturers. These, of course, do not exhaust all of the possibilities. However, we should now be able to translate most of the specifications that we may encounter.

4

Mathematical Models

Before this chapter, we have discussed the output of instruments when the input is constant. Although we have learned a considerable amount about the performance of instruments in this way, we have no information concerning the instrument's behavior when the input is a function of time. One of the problems we encounter in considering time-dependent inputs is the infinite number of inputs. Since we cannot hope to characterize the instrument response for every one of these inputs, we must seek some way to describe performance, which is independent of the input. This is accomplished by forming a theoretical model of the instrument. This model is constructed in terms of the symbols of mathematics and is referred to as the mathematical model.

We have only one goal in this chapter. After our studies, we will be able to describe the approach used to derive the mathematical model for any instrument.

In the previous chapter we considered how instruments respond to inputs that do not change with time. We shall now consider how instruments respond to inputs that are functions of time (*dynamic signals*).

Does your experience include this situation? With some reflection you will see that it does. Have you ever had your temperature "taken"? (*measured* is a better word). What does the nurse do after she puts that nasty-tasting thermometer in your mouth? Right! She leaves. Why? Right again! The thermometer must remain there long enough to respond completely to the temperature of your mouth.

Another situation with which you might be familiar often occurs when using a manometer. (If you are not familiar with this device, make one from flexible tubing and try to duplicate the event we are describing.) What happens if we suddenly change the pressure applied to a U-tube manometer? That depends upon several conditions, of course. In general, we would expect the fluid level to oscillate several times before the new level is attained. The oscillations are centered about the new level. (See Fig. 4-1.)

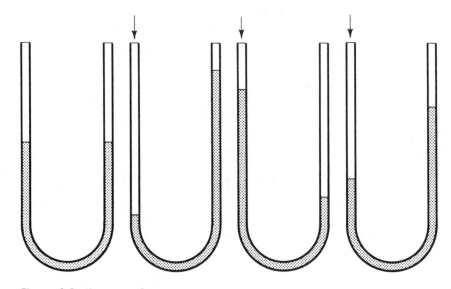

Figure 4-1. Manometer Response

Now, let us put a tubing clamp at the bottom of the U and close it part way. Now, what happens if the pressure is suddenly changed? That depends upon how tightly we clamp the tubing. With only slight restriction, we find that the oscillations are smaller and do not last as long as before. If we continue to close the clamp, we reach a point at which the oscillations cease. Further restriction increases the time required for the fluid to come to rest.

The performance of the thermometer or manometer is characteristic of that of most instruments. What measureable characteristics are available to describe this "dynamic performance"? In order to determine this we must derive the equations that govern the response of various instruments. We will refer to these equations as *mathematical models*.

4-1
The
Thermometer
The liquid-in-glass thermometer is a common and important instrument. Figure 4-2 shows a typical thermometer of this type. Note also the terms that are defined on the figure. We will use these in deriving the mathematical model of the thermometer. The input signal is temperature. You should recall from Chapter 1 that the transmission of a signal from one system to another requires that energy also be

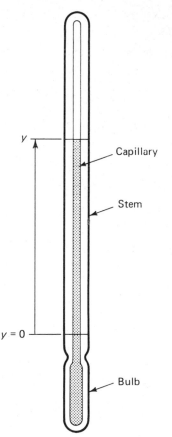

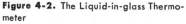

Figure 4-2. The Liquid-in-glass Thermometer

transferred. In the case of the thermometer, this energy is in the form of heat. The output signal is represented by the position of the liquid in the stem of the thermometer.

A cross section of a thermometer bulb is shown in Fig. 4-3. We will assume that the bulb is immersed in a fluid. As a result of the viscosity of the fluid, the bulb has a stagnant layer of fluid around it. The outer fluid temperature is designated T_f and that of the thermometer fluid is T_b. What is the relationship between T_f and T_b?

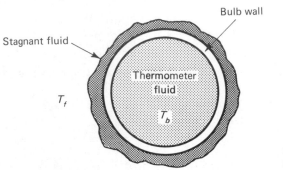

Figure 4-3. Cross Section of Thermometer Bulb

If you are familiar with the subject of heat transfer, you are aware that the heat transfer between the fluid and the thermometer fluid is given by Newton's Law of Cooling (if we ignore the presence of the glass). Thus

$$q = hA(T_f - T_b)$$

where

$q =$ heat transfer rate, energy/unit time
$h =$ film coefficient, energy/unit time-unit area-degree
$A =$ surface area of bulb

Let us assume for the present that T_f is greater than T_b. In this case, the heat will be transmitted from the fluid to the bulb. What happens to this heat? If we assume that no heat is transmitted up the stem of the thermometer, all of the energy added to the bulb is stored in the thermometer fluid. You will recall from thermodynamics that the rate of energy storage is given by

$$\frac{dE}{dt} = mc_p \frac{dT_b}{dt}$$

where

$$E = \text{energy stored}$$
$$m = \text{mass of fluid in bulb}$$
$$c_p = \text{specific heat of fluid in bulb}$$

According to our assumptions:

$$q = mc_p \frac{dT_b}{dt}$$

But

$$q = hA(T_f - T_b)$$

Thus

$$hA(T_f - T_b) = mc_p \frac{dT_b}{dt}$$

If we rearrange this slightly, we obtain

$$\frac{mc_p}{hA} \frac{dT_b}{dt} + T_b = T_f \qquad (4\text{-}1)$$

We must now eliminate T_b from Eq. (4-1). Refer to Fig. 4-2. At some temperature, $T_b = T_0$, the thermometer fluid level will be at $y = 0$. An increase of temperature will cause an increase of fluid volume. We will call this increase dV.

$$dV = \alpha V_b(T_b - T_0)$$

α is the coefficient of expansion of the fluid.

The increase in length of the filament of fluid is

$$y = \frac{dV}{A} = \frac{\alpha V_b(T_b - T_0)}{A_c} \qquad (4\text{-}2)$$

A_c is the cross sectional area of the fluid column in the thermometer stem. If we solve Eq. (4-2) for T_b, we have

$$T_b = \frac{A_c}{\alpha V_b} y + T_0$$

Substitution of the above equation into Eq. (4-1) will eliminate T_b.

$$\frac{mc_p}{hA}\frac{A_c}{\alpha V_b}\frac{dy}{dt} + \frac{A_c}{\alpha V_b}y + T_0 = T_f \qquad (4\text{-}3)$$

or

$$\frac{mc_p}{hA}\frac{dy}{dt} + y = \frac{\alpha V_b}{A_c}(T_f - T_0) \qquad (4\text{-}4)$$

T_0 appears in this equation because the bulb temperature is not nessarily zero when y is zero. This is of no particular concern since it simply indicates the position of the calibration marks, which are etched on the glass stem. For simplicity, we will drop T_0 from Eq. (4-4).

If we keep T_f constant and wait a sufficiently long time dy/dt will become zero and Eq. (4-4) reduces to

$$y = \frac{\alpha V_b}{A_c}T_f$$

This is the static input-output relationship for the thermometer. $\alpha V_b/A_c$ gives the slope of the calibration curve. This, of course, is the static sensitivity. We see, then, that the static sensitivity of the thermometer depends upon the size of the thermometer bulb, the cross-sectional area of the filament, and the thermometer fluid.

Now consider the coefficients of the derivative. Notice that three of the four terms are properties of the thermometer. We see that c_p depends upon the thermometer fluid selected, m upon both the fluid and the volume of the bulb, and A upon the bulb volume and geometry. Now, what about h, the film coefficient? A study of convective heat transfer shows that h is primarily a function of the fluid in which the thermometer is immersed, the relative velocity between the fluid and the bulb and the temperature difference, $T_f - T_b$. We see that this coefficient is a function of application and is, therefore, variable. In addition, since h depends upon ΔT, we should expect the time constant to change as ΔT changes.

4-2
The U-Tube
Manometer

The U-tube manometer is a pressure-measuring device. It is a constant-area tube bent into a U-shape and filled with liquid—usually water, mercury, or oil. If pressure is applied as shown in Fig. 4-4, the fluid will be displaced until the weight of the fluid column (h in Fig. 4-4) is equal

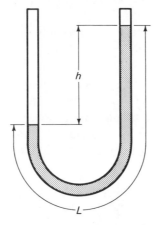

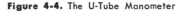

Figure 4-4. The U-Tube Manometer

to the pressure force. The input signal is pressure and the output signal is the length of the displaced fluid column, h. The transmitted energy is the work done in displacing the fluid.

Why does the manometer fluid oscillate when the pressure is changed? Let us apply Newton's Second Law of Motion and see. For our purposes, this can be stated

$$\Sigma F = ma$$

where

$$\Sigma F = \text{all external forces}$$
$$m = \text{fluid mass}$$
$$a = \text{fluid acceleration}$$

Can you identify all of these forces? Try before proceeding. Have you given this question some thought? Good! Let us begin with the external forces.

The pressure force is obviously an external force. Its value depends upon the applied pressure, p, and the cross sectional area of the manometer tube, A. In these terms, the pressure force is pA.

A second external force is the gravitational force acting on the displaced fluid column. This force is more often called weight. This is given by $\rho g h A$ where

$$\rho = \text{fluid density, slugs per cu ft}$$
$$h = \text{height of fluid column}$$
$$g = \text{gravitational acceleration}$$

The third external force is due to friction between the tube wall and the manometer fluid. To evaluate this force it is necessary to use information from the field of fluid mechanics. We will develop an expression for the fluid resistance in the next paragraphs.

To determine the shear forces between the fluid and the tube wall, it is necessary to determine the nature of the flow. For the range of pressure we would expect to encounter, the manometer fluid can be considered incompressible. In this case, the Reynolds number is indicative of the nature of the flow. This dimensionless quantity is

$$R = \frac{VD}{v}$$

where V is the fluid velocity, D the tube diameter, and v the kinematic viscosity.

Since it is difficult to determine the fluid velocity exactly, let us solve for the Reynolds number in terms of velocity. A reasonable value for tube diameter is 1/8 in. At room temperature, the kinematic viscosity of mercury is about 1.3×10^{-6} ft²/sec. Thus,

$$R = \frac{.125}{12(1.3)} V \times 10^6 = 8,000\ V$$

If the fluid is water, v is about 10^{-6} ft²/sec. With oil v is about 10^{-4} ft²/sec. We then have $R = 1000\ V$ for water and $R = 100V$ for oil.

For velocities up to 0.5 ft/sec for mercury, 4 ft/sec for water, and 40 ft/sec for oil, we can consider the flow to be laminar. As a first approximation we can usually assume the flow is laminar. For laminar flow through circular tubes, the flow is given by the Hagen-Poiseuille equation

$$Q = \frac{\Delta p \pi D^4}{128 \mu L}$$

where

$$L = \text{length of fluid column}$$
$$\Delta p = \text{pressure loss in the length, } L$$
$$Q = \text{volumetric flow rate}$$
$$\mu = \text{fluid viscosity}$$

The shear force is

$$F_s = \Delta p A = \frac{128 \mu L Q}{\pi D^4} A$$

since

$$A = \frac{\pi D^2}{4}$$

$$F_s = \frac{32\mu LQ}{D^2}$$

This equation can be written in terms of velocity if we note that

$$Q = VA = \frac{\pi D^2}{4} V$$

$$F_s = 8\pi\mu LV$$

We can write the velocity as dh/dt and the expression for shear force as

$$F_s = 8\pi\mu L \frac{dh}{dt}$$

This completes the determination of the external forces.

Noting that the mass of the manometer fluid is $AL\rho$ and its acceleration is d^2h/dt^2, we substitute in the Second Law of Motion,

$$AL\rho \frac{d^2h}{dt^2} = pA - (\rho g A)h - 8\pi\mu I \frac{dh}{dt}$$

or

$$AL\rho \frac{d^2h}{dt^2} + 8\pi\mu L \frac{dh}{dt} + (\rho g A)h = pA$$

This can be simplified somewhat by dividing by the area $(A = \pi D^2/4)$

$$L\rho \frac{d^2h}{dt^2} + \frac{32\mu L}{D^2} \frac{dh}{dt} + (\rho g)h - p \qquad (4\text{-}5)$$

The above equation is the mathematical model of the manometer. If we wait a sufficiently long time after applying a steady pressure, the derivatives of h become zero and the output h is

$$h = \frac{1}{\rho g} p$$

This indicates that the static sensitivity is $1/\rho g$.

**4-3
Mathematical
Models**

Examine the mathematical models of the two instruments Eq. (4-4) and (4-5). Are there any similarities? Both are linear differential equations. The coefficients of Eq. (4-5) are constant. By assuming the film coefficient, h, is constant, the coefficients of Eq. (4-4) can also be considered constant. For these two instruments, the mathematical models are linear differential equations with constant coefficients.

Are these models exact? That is, do they describe the performance of the instruments exactly? Since we made a number of assumptions in deriving the models, we are certain that the performance predicted by them is somewhat different from that of the instrument. The magnitude of the difference can be determined only by experiment. In developing mathematical models, we must often compromise, sacrificing accuracy in order to obtain a model for which we can obtain a solution without too much difficulty.

In the next chapter we will solve the mathematical models that describe the performance of most instruments. We will also consider solutions of the models derived in this chapter.

5

Dynamic Performance

Now that we have formulated a general mathematical model of instruments, we must consider the specific models that are useful. We must also define the dynamic characteristics that we will use to describe performance. Finally, we will consider the response to two specific dynamic inputs. We do this in detail, because the information contained in this chapter is usually used in describing the dynamic behavior of instruments.

There are many objectives in this chapter. Be sure you have mastered each before proceeding. If you have studied this chapter properly, you will be able to:

1. State the general mathematical model of instruments and reduce the general model to that for:
    (a) The zero-order instrument, (5-1)
    (b) The first-order instrument, (5-1)
    (c) The second-order instrument. (5-1)
2. Define the following terms from the coefficients of the appropriate models:
    (a) Static sensitivity, (5-1)

(b) Time constant, (5-1)

(c) Natural frequency, (5-1)

(d) Damping ratio. (5-1)

3. Identify, express mathematically, and plot each of the following input functions: (5-2)

(a) Step function,

(b) Impulse function,

(c) Terminated ramp,

(d) Sinusoidal function.

4. Derive the equations which represent the response of a first-order instrument to:

(a) The step input, (5-3)

(b) The sinusoidal input. (5-3)

5. Describe the effect of changing the time constant on the response of first-order instruments. (5-3), (5-6)

6. Derive the equations that represent the response of second-order instruments to:

(a) The step input, (5-4)

(b) The sinusoidal input. (5-4)

7. Describe the effect of changing the natural frequency and damping ratio on the response of second-order instruments. (5-4), (5-6)

In this chapter we will examine the general mathematical model of instruments and the solution of this model for various inputs. Based upon these solutions, we will be able to state the characteristics that describe the dynamic performance of instruments.

5-1
The
Mathematical
Model

Before considering the mathematical model, let us first consider what we require of our model. Naturally, we would like an equation that will describe the performance exactly. In addition, we want the equation to be simple enough that we can obtain closed solutions without undue difficulty. Unfortunately, these two requirements are not usually compatible. We are forced to compromise with a model that gives useful, but not exact, results, in order to obtain the closed solutions that are desired.

As you might assume from Chapter 4, the model that describes the performance of most instruments to a satisfactory degree is the linear differential equation with constant coefficients. If we let x represent the instrument input and y the output, the differential equation is

$$a_n \frac{d^n y}{dt^n} + a_{n-1} \frac{d^{n-1}}{dt^{n-1}} + \cdots + a_1 \frac{dy}{dt} + a_0 y = b_0 x \qquad (5\text{-}1)$$

The coefficients a_n and b_0 are physical parameters of the particular instrument we are considering. We have derived this equation for two instruments, so you should have some idea how these constants are determined.

An instrument is classified by the order of the differential equation required to describe its performance adequately. Thus, we speak of zero-order, first-order, second-order, etc., instruments. Certain measures of performance can be defined for each order of instrument. We will examine only zero-, first- and second-order responses.

Zero-order model

The *zero-order instrument* is described by the equation

$$a_0 y = b_0 x \qquad (5\text{-}2)$$

Equation (5-2) is usually written

$$y = \frac{a_0}{b_0} x \qquad (5\text{-}3)$$

In other words, the output is proportional to the input. The constant of proportionality, b_0/a_0, is the *static sensitivity*, which we will designate, K. If all of this seems familiar, it should be, since instruments behave as zero order when subjected to static inputs as discussed in Chapter 3. Since we have assumed the coefficients in Eq. (5-1) to be constant, the static sensitivity is also constant.

Some instruments behave as zero-order instruments when dynamic inputs are applied. If we apply a sinusoidal strain to a bonded strain gage, we find that the response of the gage is adequately described by Eq. (5-3) until a rather high frequency is reached. We often consider certain electronic devices to be zero order because they behave in this fashion over a wide range of frequencies.

First-order model

Not too surprisingly, the performance of the *first-order instrument* is described by

$$a_1 \frac{dy}{dt} + a_0 y = b_0 x \qquad . \qquad (5\text{-}4)$$

If Eq. (5-4) is divided by a_0, we obtain the first-order model in standard form.

$$\frac{a_1}{a_0}\frac{dy}{dt} + y = \frac{b_0}{a_0}x \qquad (5\text{-}5)$$

b_0/a_0 is the static sensitivity, K. The ratio (a_1/a_0) has time units and is called the *time constant*. We will use the symbol τ for this parameter. In terms of K and τ, the first-order model is

$$\tau\frac{dy}{dt} + y = Kx \qquad (5\text{-}6)$$

We saw in Chapter 4 that the thermometer has a first-order model. The time constant is mc_p/hA and the static sensitivity is $\alpha V_b/A_c$. Most other temperature sensors can also be modeled as first order. The parameters that are found in the time constant and the static sensitivity depend upon the particular device in use.

In the case of the thermocouple, the static sensitivity is not constant. Use of the first-order model introduces some error. Since the nonlinearity of the thermocouple is not great (see Fig. 3-15), we usually prefer to accept this error and retain our simple model.

Certain other devices may also be considered first order. Among these are a few pneumatic devices (although they really have variable static sensitivity) and some kinds of filters.

Second-order model

The second-order differential equation

$$a_2\frac{d^2y}{dt^2} + a_1\frac{dy}{dt} + a_0y = b_0x \qquad (5\text{-}7)$$

describes the performance of *second-order instruments*. Again, the static sensitivity b_0/a_0 can be formed, along with the ratio a_2/a_0 and a_1/a_0.

$$\frac{a_2}{a_0}\frac{d^2y}{dt^2} + \frac{a_1}{a_0}\frac{dy}{dt} + y = \frac{b_0}{a_0}x \qquad (5\text{-}8)$$

In a later section we will solve the second-order model. To do so, it will be necessary to solve for the roots of

$$\frac{a_2}{a_0}D^2 + \frac{a_1}{a_0}D + 1 = 0.$$

If we apply the quadratic equation we find that the roots are

$$-\frac{\dfrac{a_1}{a_0} \pm \sqrt{\left(\dfrac{a_1}{a_0}\right)^2 - 4\dfrac{a_2}{a_0}}}{2\dfrac{a_2}{a_0}}$$

This can be simplified to

$$\frac{a_1}{2a_2} \pm \sqrt{\frac{a_1^2}{4a_2^2} - \frac{a_0}{a_2}}$$

Now, let us rearrange this term as follows:

$$-\frac{a_1\sqrt{a_0}}{2\sqrt{a_2}\sqrt{a_2}\sqrt{a_0}} \pm \frac{\sqrt{a_0}}{\sqrt{a_2}}\sqrt{\frac{a_1^2}{4a_0a_2} - 1}$$

We now define the *natural frequency*

$$\omega_n = \sqrt{a_0/a_2} \qquad\qquad (5\text{-}9)$$

Using this definition, our roots are

$$-\frac{a_1\omega_n}{2\sqrt{a_0a_2}} \pm \omega_n\sqrt{\frac{a_1^2}{4a_0a_2} - 1}$$

Finally, let us define the *damping ratio*

$$\zeta = \frac{a_1}{2\sqrt{a_0a_2}} \qquad\qquad (5\text{-}10)$$

Using this definition, we write our roots

$$-\zeta\omega_n \pm \omega_n\sqrt{\zeta^2 - 1}$$

If we use the terms defined in Eqs. (5-9) and (5-10), our second-order model is

$$\frac{1}{\omega_n^2}\frac{d^2y}{dt^2} + \frac{2\zeta}{\omega_n}\frac{dy}{dt} + y = Kx \qquad\qquad (5\text{-}11)$$

Many instruments can be modeled as second order. Before listing some of these, let us refer to the manometer that we considered in Chapter 4. Look at Eq. (4-5) and the unnumbered equation that precedes it. If we compare these with Eq. (5-7) we see that $a_2 = AL\zeta$, $a_1 = 8\pi\mu L$, $a_0 = \rho gA$, and $b_0 = A$. From this we find that the natural frequency of the manometer is $\sqrt{g/l}$, the damping ratio is $(4\pi\mu/A\rho\omega_n)$, and the static sensitivity is $(1/\rho g)$.

The following instruments introduced in Chapter 1 are often considered second order: dial indicators, most accelerometers, most load cells, and pressure transducers. Sometimes the model is not obvious. Suppose we have a system as shown in Fig. 5-1. To derive the mathe-

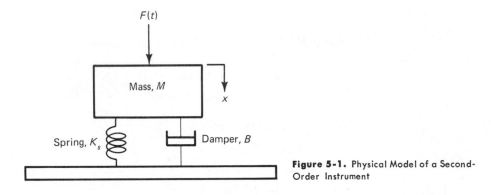

Figure 5-1. Physical Model of a Second-Order Instrument

matical model of this system, we decide that we must apply Newton's Second Law. The externally applied forces are from the spring ($F_s = K_s x$), the damper, or "dash pot", ($F_D = B\dot{x}$), and the applied force $F(t)$. We find that the mathematical model is

$$M\frac{d^2x}{dt^2} + B\frac{dx}{dt} + K_s x = F(t)$$

Or, since we prefer to express the model in the form shown in Eq. (5-11)

$$\frac{1}{\omega_n^2}\frac{d^2x}{dt^2} + \frac{2\zeta}{\omega_n}\frac{dy}{dt} + y = Kx$$

where

$$\omega_n = \sqrt{\frac{K_s}{M}}$$

$$\zeta = \frac{B}{2\sqrt{K_s M}}$$

$$K = \frac{1}{K_s}$$

Some instruments are arranged as shown in Fig. 5-1. For instance, we could make an accelerometer by suspending a mass on strain gage wire, thus making an unbonded strain gage bridge. Damping can be added by immersing the mass in some viscous fluid such as oil. Other instruments that we call second order do not look like Fig. 5-1. Suppose that we have a bourdon tube pressure gage. Where is the mass? Which part is the spring?

After looking at the instrument, you have probably decided that the tube is both mass and spring. Application of Newton's Second Law is not as simple in this case. If we are clever enough, this can be accomplished. The result is a partial differential equation. This is of course a much different mathematical model from what we have assumed. We find, for example, that the bourdon tube has many natural frequencies, and not the single natural frequency that our simplified model predicts.

If we are willing to accept some error, we can still use our second-order model. Let us assume that the bourdon tube can be modeled as shown in Fig. 5-1. The mass, M, will be the mass of the tube, and the spring constant can be measured by observing the deflection of the tube per unit force applied at the end of the tube. Now we can model the pressure gage as shown in Fig. 5-1. The natural frequency that we calculate by Eq. (5-9) will be nearly the lowest natural frequency of the instrument. This approximation is usually sufficiently accurate for most purposes. Do not forget, however, that devices such as the bourdon tube gage have many natural frequencies and not only one.

How many other devices behave as the bourdon tube pressure gage? One way to determine this is to check the device to see if the mass and spring are concentrated in one place or are distributed over the entire element. If they are distributed, our second-order model is only an approximation.

5-2
Forcing Functions

What forms can the input, x, take? This, of course, depends upon the application. In fact, virtually every conceivable form of input might occur. Fortunately, there are a few standard forms of input that give us a great deal of information concerning the performance of instruments. These are the step function, the impulse, the terminated ramp, and the sinusoidal input, each of which is discussed in the following paragraphs.

Step function

Imagine taking a thermometer, which is at equilibrium with its surroundings, and suddenly plunging it in a pot of hot water. What would a plot of input temperature versus time look like? If you sketched a curve such as that shown in Fig. 5-2, you are correct. Mathemati-

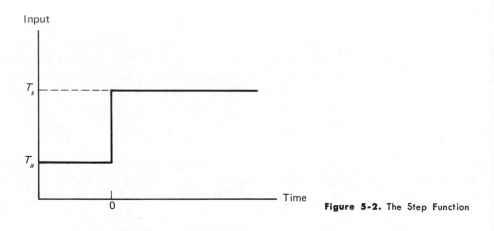

Figure 5-2. The Step Function

cally, this function is expressed

$$T = T_a \quad \text{if } t < 0$$
$$T = T_s \quad \text{if } t > 0$$

where

$$T_a = \text{initial temperature of thermometer}$$
$$T_s = \text{temperature of hot water}$$

It is often convenient to make the initial value of the input equal to zero. We can do this by defining a quantity x as follows:

$$x = T - T_a$$

With this definition, we have

$$x = T_a - T_a = 0 \ (t < 0)$$
$$x = T_s - T_a = x_s \ (t > 0)$$

(5-12)

Impulse function

In many cases the input is suddenly applied, as in the case of a step function, and, after a short time, just as suddenly removed. (Dip the thermometer into hot water and remove it quickly.) This is shown in Fig. 5-3.

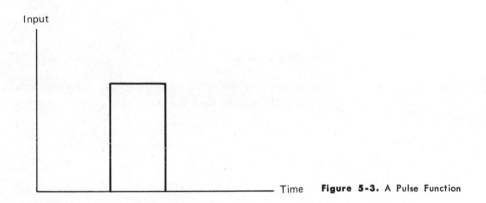

Figure 5-3. A Pulse Function

We can think of this input as two step functions superimposed. Mathematically, we have

$$x = 0 \qquad\qquad t < 0$$
$$x = x_s \qquad\qquad 0 < t < T$$
$$x = (x_s - x_s) = 0 \qquad t > T$$

We frequently encounter an input of extremely short duration, but of high magnitude. We can approximate this function as follows. Define the area of the function shown in Fig. 5-4 as A. The amplitude of the function is A/T. As T approaches 0, the amplitude increases, approaching infinity. If we call the pulse function $p(t)$, the impulse function of strength A is defined as $\lim_{T \to 0} p(t)$. If A is unity, we have the unit impulse function $I(t)$

$$I(t) = \lim_{T \to 0} p(t)$$

(5-13)

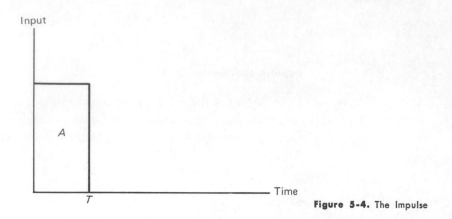

Figure 5-4. The Impulse

Terminated ramp

It is often impossible to achieve a step function exactly in a physical system. Rather, we get a function as shown in Fig. 5-5. Mathematically,

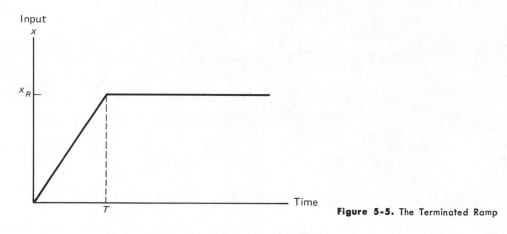

Figure 5-5. The Terminated Ramp

this function is

$$x = x_s\left(\frac{t}{T}\right) \qquad 0 < t < T$$
$$x = x_s \qquad\qquad t > T$$

(5-14)

This terminated ramp function approximates real inputs better than the step input can.

Sinusoidal function

Although not often encountered naturally, we can obtain considerable information about instrument performance by applying an input of the form

$$x = X \sin \omega t \qquad\qquad (5\text{-}15)$$

where

$$X = \text{amplitude of input}$$
$$\omega = \text{frequency, rad/sec}$$

Of the four inputs I have presented, the step and sinusoidal inputs are of particular interest at this point. On the next pages I will consider the solution of our mathematical models for first- and second-order instruments for these two inputs.

5-3
First-Order
Dynamic
Characteristics

The determination of dynamic characteristics involves the solution of linear differential equations. Can you do this? If not, refer to your differential equations text before proceeding. For convenience, let us define the differential operator

$$D = d/dt$$

and use this notation during the derivations which follow.

Step input

For the first-order instrument, the mathematical model is given by Eq. (5-6). The step input is given by Eq. (5-12). Using the differential operator, the model is

$$(\tau D + 1)y = Kx$$
$$x = 0; \ t < 0 \qquad\qquad (5\text{-}16)$$
$$x = x_s; \ t > 0$$

The solution to Eq. (5-16) is

$$y = Ce^{-t/\tau} + Kx_s$$

An appropriate initial condition is

$$y = 0 \quad \text{at} \quad t = 0$$

Applying this gives

$$0 = C + Kx_s$$

or

$$C = -Kx_s$$

so

$$y = Kx_s(1 - e^{-t/\tau}) \tag{5-17}$$

Let us see what this solution tells us about the dynamic performance of the first-order instrument. First, we should plot the response for various values of τ. Fig. 5-6 contains these curves for $\tau = 0.1$, $\tau = 0.5$, and $\tau = 1.0$ sec. What can you determine from this curve? First, it is obvious that some time is required for the output to reach its final

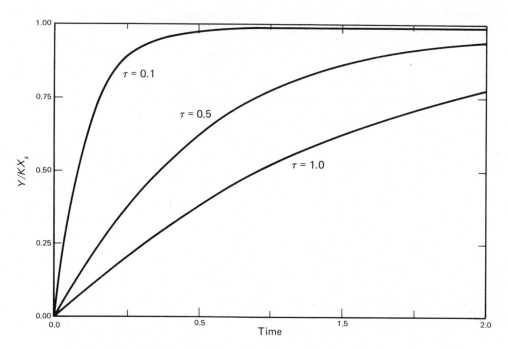

Figure 5-6. Response of a First-Order Instrument to a Step Input

value. (Remember the nurse taking your temperature?) Also, the magnitude of the time constant determines the time required for the final value to be reached. Note that $e^{-t/\tau}$ always has some value if $t < \infty$. However, for our purposes the term becomes negligibly small after a time equivalent to a few time constants.

PROBLEMS

5-1   Compute the value of Eq. (5-17) for the following values of time: $t = \tau, 2\tau, 3\tau, 4\tau, 5\tau$. (First solve Eq. (5-17) for y/Kx_s.)

5-2   What is the value of y/Kx_s at $t = 1.0$ sec if $\tau = 0.01$ sec? 0.1 sec? 1.0 sec? 10.0 sec?

Sinusoidal input

Let us now consider the response of first-order instruments to the input $x = X \sin \omega t$. Our mathematical model is

$$(\tau D + 1)y = KX \sin \omega t \qquad (5\text{-}18)$$

This solution is of the form

$$y = Ce^{-t/\tau} + AKX \sin \omega t + B\omega KX \cos \omega t \qquad (5\text{-}19)$$

where the coefficient C is determined from the initial conditions. A and B have values such that Eq. (5-19) is a solution of Eq. (5-18). Let us examine this solution under steadystate (that is, after $Ce^{-t/\tau} \simeq 0$).

We determine A and B by substituting Eq. (5-19) into Eq. (5-18). This gives

$$\tau(A\omega KX \cos \omega t - B\omega^2 KX \sin \omega t) + AKX \sin \omega t$$
$$+ B\omega KX \cos \omega t = KX \sin \omega t.$$

If Eq. (5-19) is a solution, the coefficients of $\sin \omega t$ on both sides of the equation must be equal. This must also be true for the coefficients of $\cos \omega t$.

Equating coefficients of $\sin \omega t$ gives

$$-\tau B\omega^2 + A = 1$$

Equating coefficients of cos ωt gives

$$\tau A\omega + B\omega = 0$$

We can solve these equations for A and B. The result is

$$A = \frac{1}{(1 + \tau^2\omega^2)}$$

$$B = -\frac{\tau}{(1 + \tau^2\omega^2)}$$

Our solution satisfies the differential equation for the above values of A and B. The steady-state solution of the differential equation is

$$y = \frac{KX \sin \omega t}{(1 + \tau^2\omega^2)} - \frac{\tau\omega KX \cos \omega t}{(1 + \tau^2\omega^2)}$$

This solution can be written in a more useful form by proceeding as follows. We first factor out the term $KX/\sqrt{1 + \tau^2\omega^2}$

$$y = \frac{KX}{\sqrt{1 + \tau^2\omega^2}}\left[\frac{\sin \omega t}{\sqrt{1 + \tau^2\omega^2}} - \frac{\omega\tau \cos \omega t}{\sqrt{1 + \tau^2\omega^2}}\right]$$

The bracketed term can be simplified if we observe that $1/\sqrt{1 + \tau^2\omega^2}$ is the cosine of an angle ϕ as defined in Fig. 5-7.

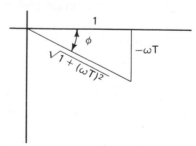

Figure 5-7. Definition of Phi

$-\omega\tau/\sqrt{1 + \tau^2\omega^2}$ is the sine of ϕ. Thus, we can write

$$y = \frac{KX}{\sqrt{1 + \tau^2\omega^2}}(\sin \omega t \cos \phi + \cos \omega t \sin \phi)$$

or

$$y = \frac{KX}{\sqrt{1 + \tau^2\omega^2}} \sin(\omega t + \phi) \qquad (5\text{-}20)$$

where

$$\phi = \tan^{-1}(-\omega\tau)$$

Now, let us look at a plot of the input $KX \sin \omega t$ versus time and output y versus time for a given value of ω and τ (Fig. 5-8). We see

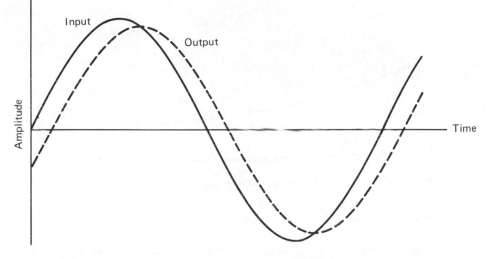

Figure 5-8. Input and Output for a First-Order Instrument

that the maximum value of y is less than the maximum value of KX. In addition, these peak values do not occur at the same time. The output peak occurs ϕ/ω time units later than the input.

We can see how the output varies with frequency by plotting the amplitude ratio, Y/KX, versus ω and ϕ versus ω. This is done in Fig. 5-9 for $\tau = 1$ and 0.1 sec. From these curves we see that the ratio of amplitudes (Y/KX) decreases with increasing frequency and that the phase angle (ϕ) also decreases (becomes more negative) with increasing frequency. As the frequency approaches infinity, ϕ approaches $-90°$. Note also that increasing τ causes the phase angle and amplitude ratio to decrease more rapidly.

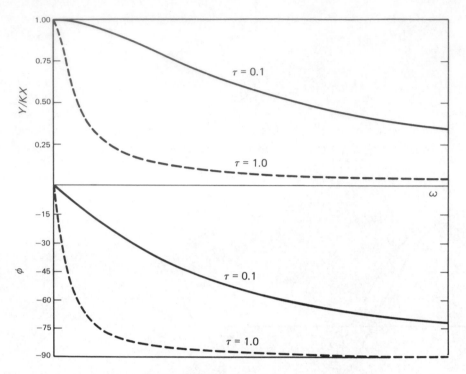

Figure 5-9. First-Order Frequency Response Curves

The time constant

Examine Figs. 5-6, 5-8, and 5-9. From these we can see that the first-order instrument begins to respond as soon as an input signal is imposed, but that a considerable amount of time is required for the output to reach its final or steady state value. If the input is not constant but continues to change, the output of the instrument is always in error. The dynamic characteristic, time constant, indicates the speed with which an instrument will respond. Instruments with small time constants respond more rapidly than those with large time constants.

From Fig. 5-9 we can see that for signals similar to a sinusoid (that is, periodic) the output signal is always lower in amplitude than the input signal. In addition, the output signal lags behind the input signal. For a constant value of τ, the attenuation and lag increase with increasing frequency. If the frequency is constant, the attenuation and lag increase with increasing values of time constant.

Let us return to the liquid-in-glass thermometer we discussed in

Chapter 4. By now it should be rather obvious that the length of time the nurse must leave that thermometer in your mouth is determined by its time constant. From Prob. 5-1 we see that a time equivalent to about five time constants must elapse for the output to reach 99 percent of its final value. For instance, if the time constant is about 30 sec, she should wait $5 \times 30 = 150$ sec or almost three minutes to be sure of obtaining a good indication of the temperature of your mouth.

PROBLEMS

5-3   A thermocouple with a time constant of 0.3 sec and a static sensitivity of 0.025 mV/°F is suddenly immersed in a bath of hot oil which is at 350°F. The initial temperature of the thermocouple measuring and reference junctions was 70°F. What is the output of the thermocouple at the following values of time: $t = 0.15$ sec, $t = 0.30$ sec, $t = 1.0$ sec?

5-4   It is necessary to select a liquid-in-glass thermometer for a particular application. The specification for the application requires that the thermometer response to a step input must be such that $y/Kx_s = 0.95$ within 120 sec. What is the required value of the time constant?

5-5   A first-order instrument is subjected to a step input. After 3 sec, $y/KX_s = 0.75$. What is the time constant of this instrument?

5-6   A temperature-sensing device that is first order has a time constant of 0.75 sec and a static sensitivity of 1.2 mV/°F. A sinusoidal temperature shown in Fig. P5-6 is imposed. Assuming that this input has been present long enough so that steady state conditions prevail,

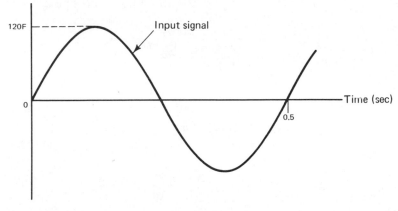

Figure P5-6.

what is the output from the instrument at the beginning of an input
cycle (marked $t = 0$ on the figure)? at $t = 0.1$? at $t = 0.2$?

5-7   The input to a first-order instrument is a terminated ramp (Fig. 5-5).
Solve Eq. 5-6 for this input and plot the results. (Plot y/Kx_R as a function of time.)

5-8   A first-order instrument has the input $x = 10 \sin \omega t$. The static sensitivity is 2.5 output units per input unit. The time constant is 0.2 sec.
Compute the phase angle and maximum value of y for $\omega = 0.1, 1, 10$
and 100 rad/sec.

5-9   What is the value of ϕ and y/KX for $\omega = 10$ rad/sec for each of the
following values of time constant: $\tau = 0.1, 1, 10, 100$ sec?

5-4
Second-Order
Dynamic
Characteristics

We will now solve the second-order model as given in Eq. (5-11) for
a step input and a sinusoidal input. The influence of the dynamic
characteristics, natural frequency and damping ratio, which were
defined by Eqs. (5-9) and (5-10), can be seen from these solutions.

The step input

The equation we wish to solve is obtained by substituting the step
input defined by Eq. (5-12), for x in Eq. (5-11).

$$\frac{1}{\omega_n^2} \frac{d^2y}{dt^2} + \frac{2\zeta}{\omega_n} \frac{dy}{dt} + y = Kx$$

$$x = 0 \qquad t < 0$$

$$x = x_s \qquad t > 0$$

We will again use the differential operator notation, $d/dt = D$

$$\left(\frac{1}{\omega_n^2} D + \frac{2\zeta}{\omega_n} D + 1\right)y = Kx_s \tag{5-21}$$

$$t > 0$$

You probably recall from differential equations that the solution of
equations such as Eq. (5-21) has two parts. One part (often called the
complimentary or *homogeneous* solution) is the solution to

$$\left(\frac{1}{\omega_n^2} D + \frac{2\zeta}{\omega_n} D + 1\right) = 0$$

The other part (usually called the *particular solution* or the *particular integral*) is any solution, y, which satisfies Eq. (5-21).

The complimentary solution depends upon the roots of the characteristic equation. In our case this equation is

$$D^2 + 2\zeta\omega_n + \omega_n = 0$$

Application of the quadratic equation yields the roots

$$R_1, R_2 = -\zeta\omega_n \pm \omega_n\sqrt{\zeta^2 - 1}$$

At this point we can see some of the influence of the damping ratio. If ζ is greater than one, the roots are real and not repeated. For ζ equals one, the roots are real and repeated. Finally, if ζ is less than one, the roots are imaginary. These roots give three distinct forms of solution. Let us look at them.

First, for $\zeta > 1$ the roots are $R_1 = -\omega_n(\zeta + \sqrt{\zeta^2 - 1})$ and $R_2 = -\omega_n(\zeta - \sqrt{\zeta^2 - 1})$. Thus, the complimentary solution is

$$y_c = C_1 e^{-(\zeta + \sqrt{\zeta^2-1})\omega_n t} + C_2 e^{-(\zeta - \sqrt{\zeta^2-1})\omega_n t}$$

We can see by inspection that the particular solution is

$$y_{pi} = Kx_s.$$

Thus,

$$y = y_c + y_{pi} = C_1 e^{-(\zeta + \sqrt{\zeta^2-1})\omega_n t} + C_2 e^{-(\zeta - \sqrt{\zeta^2-1})\omega_n t} + Kx_s$$

If we take the initial conditions as $y = 0$ and $dy/dt = 0$, when $t = 0$,

$$C_1 = \frac{Kx_s(\zeta - \sqrt{\zeta^2 - 1})}{2\sqrt{\zeta^2 - 1}}$$

$$C_2 = \frac{-Kx_s(\zeta + \sqrt{\zeta^2 - 1})}{2\sqrt{\zeta^2 - 1}}$$

The complete solution is

$$\frac{y}{Kx_s} = \frac{\zeta - \sqrt{\zeta^2 - 1}}{2\sqrt{\zeta^2 - 1}} e^{-(\zeta + \sqrt{\zeta^2-1})\omega_n t}$$

$$-\frac{\zeta + \sqrt{\zeta^2 - 1}}{2\sqrt{\zeta^2 - 1}} e^{-(\zeta - \sqrt{\zeta^2-1})\omega_n t} + 1 \tag{5-22}$$

From this solution, we see that the output approaches the value Kx_s exponentially. For a fixed value of ω_n, increasing the value of ζ increases the time required for y to approach its final value.

Consider the case for which $\zeta = 1$. We now have repeated roots $R_1 = R_2 = -\omega_n$. The solution is

$$y = C_1 e^{-\omega_n t} + C_2 t e^{-\omega_n t} + Kx_s \qquad (5\text{-}23)$$

From the initial conditions

$$C_1 = -Kx_s$$
$$C_2 = -K\omega_n x_s.$$

Substituting these values in Eq. (5-23) gives

$$y/Kx_s = 1 - (1 + \omega_n t)e^{-\omega_n t} \qquad (5\text{-}24)$$

Once again, the output approaches Kx_s exponentially. The rate at which this occurs is determined by the value of the natural frequency, ω_n.

The final case is for damping ratios less than one. In this case, it is convenient to write the roots as

$$R_1, R_2 = -\zeta\omega_n \pm i\omega_n\sqrt{1 - \zeta^2}$$

For imaginary roots, the solution is

$$y = Ce^{-\zeta\omega_n t} \sin\left(\sqrt{1 - \zeta^2}\,\omega_n t + \phi\right) + Kx_s$$

Applying the initial conditions, we have at $t = 0$,

$$y = C \sin \phi + Kx_s = 0$$

and

$$dy/dt = -\zeta\omega_n C \sin \phi + C(\omega_n\sqrt{1 - \zeta^2}) \cos \phi = 0$$

From these equations, we have

$$\phi = \sin^{-1}\left(\sqrt{1 - \zeta^2}\right)$$

$$C = -\frac{Kx_s}{\sqrt{1 - \zeta^2}}$$

The solution then is

$$\frac{y}{Kx_s} = 1 - \frac{e^{-\zeta\omega_n t}}{\sqrt{1 - \zeta^2}} \sin\left(\sqrt{1 - \zeta^2}\,\omega_n t + \phi\right)$$

$$\phi = \sin^{-1}\left(\sqrt{1 - \zeta^2}\right)$$

(5-25)

Note that this solution is quite different from the previous two. In the case of $\zeta < 1$, the solution oscillates about the final value Kx_s for some time before settling out (ceasing to oscillate). This solution, along with the solutions for $\zeta > 1$ and $\zeta = 1$ are plotted in Fig. 5-10.

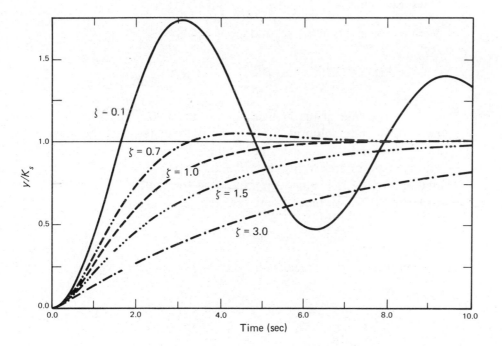

Figure 5-10. Second-Order Step Response

The case $\zeta = 1$ represents a division between the oscillatory and non-oscillatory response. We say that the instrument is critically damped when the damping ratio is unity. If $\zeta < 1$ the instrument is under-damped. The instrument is overdamped for values of damping ratio greater than one.

PROBLEMS

5-10 A pressure transducer has a natural frequency of 30 rad/sec, damping
ratio of 0.1 and static sensitivity of 8 mV/psi. A step pressure input
of 100 psi is applied. What is the value of the output when $t = 0.05$
sec? When $t = 0.5$ sec? When $t = 1$ sec?

5-11 Repeat Prob. 5-10 for $\zeta = 0.6$, $\zeta = 1.0$, and $\zeta = 2.0$.

5-12 How much time must elapse in order that the output of the instrument will remain within $\pm 5\%$ of its final (steady state) value?

5-13 Repeat Prob. 5-12 for $\zeta = 0.6$ and $\zeta = 1.0$.

Sinusoidal input

With a sinusoidal input our model is

$$\frac{1}{\omega_n^2} \frac{d^2y}{dt^2} + \frac{2\zeta}{\omega_n} \frac{dy}{dt} + y = KX \sin \omega t. \qquad (5\text{-}26)$$

As in the case of the first-order instrument we are concerned with
the steady state solution. Since y_c contains a term of the form e^{-t},
this part of the solution becomes negligibly small after some finite time
and is of no interest to us. The remainder of the solution, y_{pi}, can be
determined by the same technique used for the first-order model. In
order to do so, we first assume that the solution is

$$y = AKX \sin \omega t + B\omega KX \cos \omega t$$

Substituting this assumed solution in Eq. (5-26) gives

$$\frac{1}{\omega_n^2}(-A\omega^2 KX \sin \omega t - B\omega^3 KX \cos \omega t)$$

$$+ \frac{2\zeta}{\omega_n}(A\omega KX \cos \omega t - B\omega^2 KX \sin \omega t)$$

$$+ (AKX \sin \omega t + B\omega KX \cos \omega t) = KX \sin \omega t$$

The next step is to equate the coefficients of like functions of time
on each side of the equation. Thus,

$$-A\left(\frac{\omega^2}{\omega_n^2}\right) - B\frac{2\zeta\omega^2}{\omega_n} + A = 1$$

and

$$-B\left(\frac{\omega^3}{\omega_n^2}\right) + A\frac{2\zeta\omega}{\omega_n} + B\omega = 0$$

or

$$A\left(1 - \frac{\omega^2}{\omega_n^2}\right) - B\omega\left(\frac{2\zeta\omega}{\omega_n}\right) = 1$$

$$B\omega\left(1 - \frac{\omega^2}{\omega_n^2}\right) + A\left(\frac{2\zeta\omega}{\omega_n}\right) = 0.$$

To make it easier to write the equations, let us define

$$M = \left(1 - \frac{\omega^2}{\omega_n^2}\right)$$

$$N = \frac{2\zeta\omega}{\omega_n}$$

If we make this substitution and solve for A and B, we obtain

$$A = \frac{M}{M^2 + N^2} \quad \text{and} \quad B = \frac{-N}{\omega(M^2 + N^2)}$$

The steady state solution is

$$y = \frac{MKX \sin \omega t}{(M^2 + N^2)} - \frac{MKX \cos \omega t}{(M^2 + N^2)}$$

This solution is perfectly correct, but we can simplify it somewhat as follows:

$$y = \frac{KX}{\sqrt{M^2 + N^2}}\left[\frac{M \sin \omega t}{\sqrt{M^2 + N^2}} - \frac{N \cos \omega t}{\sqrt{M^2 + N^2}}\right] \qquad (5\text{-}27)$$

Referring to Fig. 5-11, we see that

$$\frac{M}{\sqrt{M^2 + N^2}} = \cos \phi$$

and

$$\frac{-N}{\sqrt{M^2 + N^2}} = \sin \phi$$

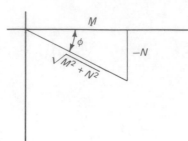

Figure 5-11. Definition of Phi

If these terms are introduced into Eq. (5-27), we have

$$y = \frac{KX}{\sqrt{M^2 + N^2}}[\sin \omega t \cos \phi + \cos \omega t \sin \phi]$$

or

$$y = \frac{KX}{\sqrt{M^2 + N^2}} \sin (\omega t + \phi)$$

where

$$\phi = \tan^{-1} \frac{-N}{M}$$

Finally, recalling the definitions of M and N, we have

$$y = \frac{KX}{\sqrt{\left(1 - \frac{\omega^2}{\omega_n^2}\right)^2 + \left(\frac{2\zeta\omega}{\omega_n}\right)^2}} \sin (\omega t + \phi) \qquad (5\text{-}28)$$

$$\phi = \tan^{-1} \frac{-2\zeta\omega/\omega_n}{\left(1 - \frac{\omega^2}{\omega_n^2}\right)}$$

Now, what can we say about the response of second-order instruments to a sinusoidal input? Consider the curves that are plotted in Figs. 5-12 and 5-13. The first set of curves is the maximum value of Y/KX versus ω/ω_n for various values of the damping ratio. Note that the amplitude of the output exceeds, that of the input in the neighborhood of the natural frequency, for sufficiently low values of the damping ratio. This phenomenon is called *resonance* and the frequency at which the maximum amplitude occurs is called the *resonant frequency*. For $\zeta = 0$, the resonant frequency is the natural frequency.

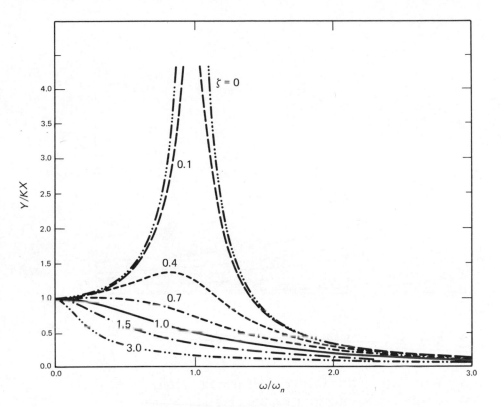

Figure 5-12. Second-Order Frequency Response-Magnitude Ratio

The second set of curves is phase angle, ϕ, plotted as a function of ω/ω_n for various damping ratios. Note that the maximum phase shift is $-180°$ and that $\phi = -90°$ for $\omega/\omega_n = 1$.

PROBLEMS

5-14   A second-order instrument is subjected to a sinusoidal input. The natural frequency is 20 rad/sec. Compute the amplitude ratio and phase angle for an input frequency of 10 rad/sec at $\zeta = 0, \zeta = 0.5$, $\zeta = 1.0$, and $\zeta = 2.0$.

5-15   Repeat Prob. 5-14 for an input frequency of 20 rad/sec.

5-16   Repeat Prob. 5-14 for an input frequency of 100 rad/sec.

5-17   A second-order instrument with a damping ratio of 0.5 is subjected to a sinusoidal input of 10 rad/sec. Compute the amplitude ratio for $\omega_n = 1$ rad/sec, 10 rad/sec and 100 rad/sec.

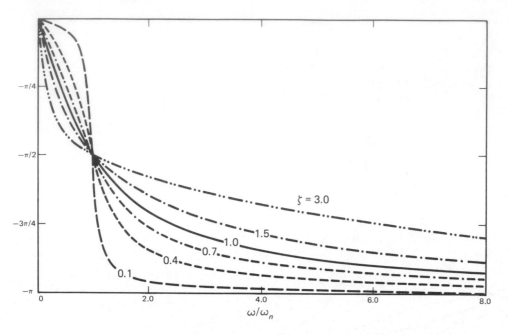

Figure 5-13. Second-Order Frequency Response-Phase Angle

5-5
The Importance
of Frequency
Response

We should be sure that we are thoroughly familiar with the frequency response (response to a sinusoidal input) of first- and second-order instruments. This information is very useful in visualizing and predicting the way an instrument will respond to any input. We are able to do this because most inputs we encounter can be expressed as some combination of sinusoidal terms. We will develop this idea more completely in Chapter 6. Be sure to learn Eq. (5-20) and Eq. (5-28) and be able to reproduce Figs. 5-9, 5-12, and 5-13. We will use this information often in the next chapter.

5-6
Dynamic
Characteristics

Three dynamic characteristics are of interest to us. These are the time constant (τ), the natural frequency (ω_n), and the damping ratio (ζ). Each of these are defined in terms of the coefficients of the appropriate mathematical model. These coefficients represent parameters of the physical model (that is, of the instrument). If we have derived the mathematical model for a particular device (as we did in Chapter 4) we know which variables are included in the various dynamic characteristics.

It is important to know how variation of the dynamic characteristics alters the dynamic behavior of an instrument. Let us consider the first-order instrument. The dynamic characteristic that is of interest is the time constant. Examine Eq. (5-17) and Fig. 5-6. How can we make a first-order instrument respond more rapidly? From the figure we note that decreasing the time constant increases the speed of response. If $\tau = 0$, Eq. (5-17) becomes $y = Kx_s$. This is the zero-order instrument in which the output is always proportional to the input. We usually try to make the time constant as small as possible, although it cannot be reduced to zero.

Figure 5-9 and Eq. (5-20) show the influence of the time constant on the first-order frequency response. Since an increase of τ slows the response, we would expect the amplitude ratio to become smaller and the phase angle to become more negative as τ becomes larger. This is clear in Fig. 5-9.

Two characteristics govern the dynamic response of second-order instruments—the natural frequency and the damping ratio. We will consider the effect of the damping ratio first. When the step response was considered, we found that the damping ratio determined whether the output of the instrument would oscillate. Damping ratios that are less than one indicate the presence of oscillations. This characteristic also controls the rate at which these oscillations die out (decay). If $\zeta = 0$, oscillations continue indefinitely.

Frequency response is also affected by the damping ratio. For $0 \leq \zeta < 1$, resonance occurs. For $\zeta \geq 1$, it does not. You may wonder what value of ζ one should choose (if one can choose). Look at Fig. 5-12. We would like for Y/KX to be as close to one as possible over the largest possible range of input frequency. This occurs for values of damping ratio between 0.6 and 0.7.

Although its effect is not as obvious, the natural frequency also has considerable effect on dynamic response. From Eq. (5-25), we see that for a fixed value of ζ, an increase of ω_n will decrease the time required for the oscillations to decay. An increased natural frequency will also increase the range of input frequency that the instrument is capable of responding to adequately.

In general, we choose an instrument with a high natural frequency and a damping ratio of 0.6–0.7. Of course, "high" is a relative term. If the maximum expected input frequency is one rad/sec, $\omega_n = 100$ rad/sec is "high." If the anticipated input frequency is 1000 rad/sec, $\omega_n = 100$ rad/sec is probably inadequate. In many cases we cannot obtain our desired value of ζ, but may find values ranging down to $\zeta = 0.1$ or less. Unless ω_n is rather large, the frequency range over which we can use the instrument is severely restricted.

**5-7
Examples of
Specifications of
Second-Order
Characteristics**

We have indicated the appropriate mathematical model for most of the examples we have been discussing. Let us now look at how the dynamic characteristics are specified by manufacturers.

Temperature-sensing devices

Since these are first-order instruments, we would expect to find the time constant specified. If, however, you look at a few specifications of these instruments, you will find that τ is not usually specified. It is not too difficult to determine the reason for this. Look at Eq. (4-4). The coefficient of dy/dt in this equation is the time constant. Note that one of the terms in the time constant is not a property of the thermometer. The film coefficient, h, is dependent upon the properties of the fluid in which the instrument is immersed, as well as the fluid velocity and temperature. Because of this, there is no way to specify a time constant unless the specific application is known.

You can establish relative values of the time constants of several instruments. We know that $\tau = mc_p/hA$. If we assume all thermometers will be used in the same way, h can be treated as a constant and we can estimate values of $\tau h = mc_p/A$. The three parameters m, c_p, and A are properties of the thermometers and can be estimated or measured.

Accelerometers

In Chapter 3 we discussed static characteristics for an accelerometer. The following are dynamic characteristics for the same instrument.

Approximate Natural Frequency	400 Hz
Damping	0.7 ± 0.15 of critical at room temperature
Vibration	to 300 g (50 to 2500 Hz)
Shock	75 g—15 ms

Since the instrument is modeled as a second-order device we would expect the natural frequency and damping ratio to be specified. These are 400 Hz and 0.7 ± 0.15, respectively. Notice that the "approximate natural frequency" is specified. This means that the instrument is designed for this value of ω_n, but because of manufacturing and material variations, the natural frequencies of individual instruments will vary somewhat. The manufacturer usually measures this value for each instrument and furnishes it for the purchaser.

Although we will not prove it here, small accelerations applied rapidly can cause more damage than large slowly applied accelerations. Two specifications, shock and vibration, indicate the maximum transient and oscillatory accelerations that the instrument can withstand without sustaining damage. Shock is the acceleration that occurs when objects collide. According to our specification, the accelerometer is capable of accepting accelerations of 75 g, which are applied within a 15 ms time interval. Figure 5-14 shows the time history of such a

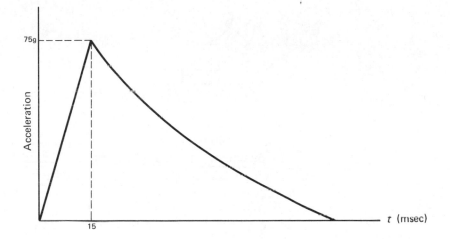

Figure 5-14. A Shock Input

shock. *Vibration* refers to the accelerations that occur because of oscillatory motion. This instrument will not be damaged by vibration of 30 g amplitude over the range 50–2500 Hz.

Load cell

We also looked at load cell specifications in Chapter 3. The specifications presented there were complete except for one, which is

Spring rate 2.5×10^6 lb/in.

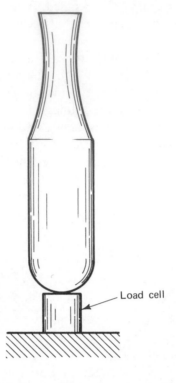

Figure 5-15. An Application of the Load Cell

Why did the manufacturer not specify natural frequency and damping ratio? Consider a typical application of a load cell. In Fig. 5-15 a load cell is being used to measure the thrust of a rocket engine. The mass of the rocket is much greater than that of the load cell. Now, the natural frequency is $\sqrt{K_s/M}$. K_s is the spring rate of the load cell. What about M? It is certainly not the mass of the cell. We must use the rocket's mass. Since the manufacturer cannot know your particular application, he is unable to specify ω_n. Specification of the damping ratio is also impossible because mass is involved.

Other specifications

Sometimes information concerning dynamic performance is given in terms other than ω_n and ζ. For example, a manufacturer may provide a frequency response curve as part of his specifications. In some cases he might give a written description of this curve, stating the range in which the amplitude ratio will lie for a given range of input frequency.

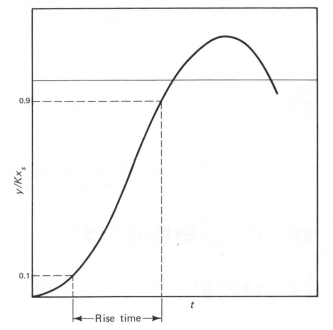

Figure 5-16. Rise Time

This sort of description is most often used in describing amplifiers and various recording devices.

Rise time is also used to describe dynamic performance. This term describes the response of a second-order instrument. As shown in Fig. 5-16, the rise time is defined as the time required for the output to rise from 10 percent to 90 percent of its final value. Note that this specification does not indicate the magnitude or duration of any oscillation that might occur.

There are other ways in which dynamic response is described, but we will not describe any of them now. You should be able to apply your knowledge of the mathematical model and its solutions to determine the meaning of unfamiliar specifications.

6

Complex Instruments and General Inputs

By now, you should be pretty good at describing and predicting the response of first- and second-order instruments to constant inputs and to step and sinusoidal inputs. It would be pleasant if we could stop at this point. Unfortunately, however, only the very simplest of instruments can be categorized as first or second order. What do we do in those cases in which the models that have been developed are not suitable? This is one of the topics that will be discussed in this chapter.

A second problem becomes obvious as soon as we consider real input signals. The step and sinusoidal inputs occur rarely, if at all. More often we have much less tractable inputs to deal with. Can we apply what we have learned about instrument performance to these inputs? We will see how this is done in the last half of this chapter.

Once again, there are a number of objectives. If you master each in sequence, you will be able to:

1. Calculate the static sensitivity and error for several first- and second-order instruments connected in series. (6-2)

2. Determine the response of a complex instrument by utilizing the response of its component parts. (6-3)
3. Express the frequency response of higher-order instruments in terms of the frequency response of first- and second-order instruments. (6-3)
4. Make logarithmic plots of the frequency response of any instrument. (6-3)
5. Describe the characteristics of periodic, transient, and random signals. (6-4)
6. Demonstrate the method for predicting the response of instruments to
    (a) periodic signals (6-4)
    (b) transient signals (6-4)
    (c) random signals (6-4)

6-1
Higher-Order
Instruments

Only the most simple and elementary instruments can be considered zero, first, or second order. In most cases we must assemble a system composed of a number of first- and second-order components to perform a particular measurement. The system shown in Fig. 6-1 is an example of a higher-order instrument. Each of the components (LVDT and digital readout) has static and dynamic characteristics as discussed in the previous chapters. For other cases, the instrument may have a third- or higher-order mathematical model. Our task now is to learn how we can apply the information we have from Chapters 3 and 5 to these situations.

Before we consider how we might predict the performance of higher-

Figure 6-1. A Higher-Order Instrument
(Courtesy Scaevitz Engineering)

order instruments, let us make a distinction between some of the characteristics we have studied. Think for a moment how static characteristics are determined. All of these characteristics (with the possible exception of the static sensitivity) must be determined by static calibration. The dynamic characteristics can at least be estimated analytically as we did in Chapter 4. Static characteristics are treated as if they were random, whereas dynamic characteristics are treated as if they were deterministic. We must remember this difference as we continue our discussion of higher-order instruments.

It is possible to consider higher-order instruments in either of two ways. First, we may attempt to reduce the instrument to a number of simpler components. Another approach is to consider the system as a whole. The first approach is better when considering dynamic performance. We would usually employ the second when considering static performance.

6-2
Static Response of Higher-Order Instruments

It is usually best to determine the static response of a complex instrument directly by calibrating the entire system. However, you may find it necessary to estimate some of the static characteristics of a system before it is assembled. In this case we consider the system to be a number of first- and second-order components connected in series as shown in Fig. 6-2. Under static conditions (see Chapter 3), the output of any of the components is the input to the component multiplied by its static sensitivity,

$$y = Kx \tag{6-1}$$

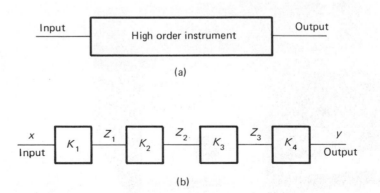

Figure 6-2. A Complex Instrument (a) and its Component Parts (b)

What is the value of the static sensitivity of the instrument shown in Fig. 6-2? We can write an expression for K in terms of K_1, K_2, K_3, and K_4 without any difficulty.

$$z_1 = K_1 x$$

$$z_2 = K_2 z_1$$

$$z_3 = K_3 z_2$$

$$y = K_4 z_3$$

When we eliminate the z's from these equations, we find that

$$y = K_1 K_2 K_3 K_4 x$$

The static sensitivity of the instrument then is

$$K = K_1 K_2 K_3 K_4 \tag{6-2}$$

This statement is correct for any number of elements connected in series. It is also true if the static sensitivity is a function of input rather than a constant.

You will recall (from Chapter 3) that the output from each component is subject to some error, and that this error is randomly distributed. This error is usually specified by the manufacturer of the component. If we combine components as in Fig. 6-2, what error should we state for the output of the composite instrument?

Let us be a little more specific. Suppose the input and output of the instrument shown in Fig. 6-2 have the same units and that the error in each component is ± 1 unit. Will the error in the output be the sum of the component errors (± 4 units)? That really does not seem reasonable. Remember that we decided that the error is randomly distributed. It is also safe for us to assume that the error in a particular component is independent of those of the other components. In this situation it is not likely that all the errors will bear the same sign. You will agree that stating the error of a composite instrument as the sum of the errors of its components is not correct. The result is a predicted error that is larger than any we are likely to obtain.

If we cannot simply sum component errors to determine the overall error, what should we do? In considering the cause of the random error in output, we observe that it occurs because of random variation in K, the static sensitivity. Thus, when examining the error of the composite instrument, we see that the random variables are the component static sensitivities. Equation (2-13) shows how we calculate the standard deviation of a function of two or more random variables.

$$\sigma_K = \left[\left(\frac{\partial K}{\partial K_1} \right)^2 \sigma_{K_1}^2 + \left(\frac{\partial K}{\partial K_2} \right)^2 \sigma_{K_2}^2 + \cdots \right]^{1/2}$$

For the instrument we are considering

$$K = K_1 K_2 K_3 K_4$$

$$\frac{\partial K}{\partial K_1} = K_2 K_3 K_4$$

$$\frac{\partial K}{\partial K_2} = K_1 K_3 K_4$$

$$\frac{\partial K}{\partial K_3} = K_1 K_2 K_4$$

$$\frac{\partial K}{\partial K_4} = K_1 K_2 K_3 \tag{6-2}$$

Perhaps we should consider this result more closely. σ_K depends upon the value of static sensitivities of the components as well as their errors. This leads us to an interesting point. If we wish to reduce σ_K, we can do so by decreasing one or more of the static sensitivities or by attempting to decrease the error of one or more of the components.

If we choose to decrease only one value of error, which one should we select? That depends upon the values of the products $K_1 K_2 K_3$, $K_1 K_2 K_4$, $K_1 K_3 K_4$, and $K_2 K_3 K_4$. Suppose that

$$K_1 K_2 K_3 = 0.001$$

$$K_1 K_2 K_4 = 0.01$$

$$K_1 K_3 K_4 = 0.1$$

$$K_2 K_3 K_4 = 1.0$$

$$\sigma_{K_1} = \sigma_{K_2} = \sigma_{K_3} = 0.01$$

From Eq. (2-13)

$$\sigma_K = [(K_2 K_3 K_4)^2 \sigma_{K_1}^2 + (K_1 K_3 K_4)^2 \sigma_{K_2}^2 + (K_1 K_2 K_4)^2 \sigma_{K_3}^2 + (K_1 K_2 K_3)^2 \sigma_{K_4}^2]^{1/2}$$

$$\sigma_K = [(1.0)^2 (0.01)^2 + (.1)^2 (0.01)^2 + (0.01)^2 (0.01)^2 + (0.001)^2 (0.01)^2]^{1/2}$$

$$= .01 \sqrt{1.010101}$$

Now, let us see what happens when we halve σ_{K_1}

$$\sigma_K = 0.01\sqrt{0.260101}$$

If σ_{K_2} is halved,

$$\sigma_K = 0.01\sqrt{1.000351}$$

If σ_{K_3} is halved,

$$\sigma_K = 0.01\sqrt{1.0100035}$$

and if σ_{K_4} is halved,

$$\sigma_K = 0.01\sqrt{1.01010025}$$

Notice that only by changing σ_{K_1} do we gain much accuracy.

Prediction of the other static characteristics from component characteristics can become complicated and imprecise. If you are concerned about the linearity, resolution, hysteresis, and so on, of a particular combination, the best procedure is to perform a static calibration as described in Chapter 3.

**6-3
Dynamic
Response of
Higher-Order
Instruments**

One means of predicting the dynamic performance of a higher-order instrument is to follow the procedure described in Chapters 4 and 5. That is, we formulate an appropriate model of the instrument and from the model derive the differential equation that describes the performance of the instrument. We can then solve this equation for various inputs.

Although the above procedure can be expected to yield excellent results, it requires knowledge involving the design of various parts of the instrument. Quite often this information is not available to us. It is desirable to develop a procedure that requires only knowledge concerning dynamic performance that is readily available. Included are such items as time constant, static sensitivity, natural frequency, and damping ratio. This procedure consists of reducing the instrument to its component parts, writing the mathematical model for each, and finally, determining the response of each component to the particular input imposed upon it. We will develop these ideas in the following paragraphs.

Reduction to components

We can simplify a complex instrument by considering its component parts. One means of accomplishing this is to reduce the instrument to its functional elements as we did in Chapter 1. Each of these will probably be a simple first- or second-order device. The blocks in Fig. 6-2(b) thus might represent the functional elements of an instrument.

Now that we have divided the instrument into its component parts, is it possible to construct the differential equations that describe the performance of the components? Suppose that we have the dynamic characteristics of each of the components. Can we write the appropriate mathematical model? Let us see if we can.

PROBLEMS

6-1   The static sensitivity of a particular instrument is 100 output units per input unit and the time constant is 0.01 sec. What is the mathematical model for this instrument?

6-2   The static sensitivity of an instrument is 10 output units per input unit. Its natural frequency is 300 rad/sec and the damping ratio is 0.7. What is the mathematical model of the instrument?

Now that you have recalled how to write the mathematical model of each of the components if you are given the dynamic characteristics, let us see how to use them. Consider an instrument which is composed of one first-order element and one second-order element connected in series as shown in Fig. 6-3. Suppose $x(t)$ is a step change of input. What is y? We shall begin by determining x'. From Chapter 5 we learned that the response of a first-order instrument to a step change of input is

$$x' = Kx_s(1 - e^{-t/\tau})$$

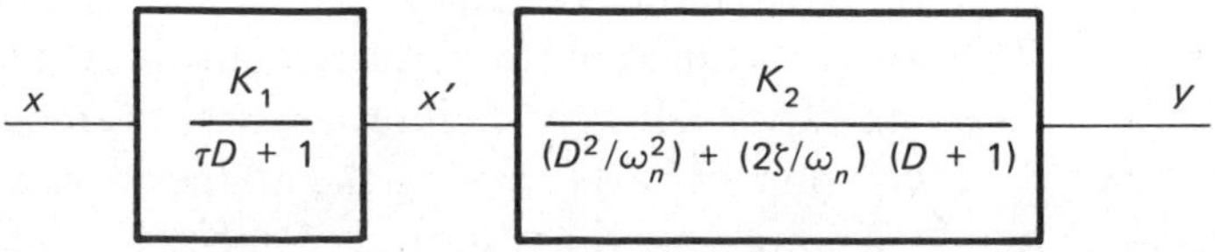

Figure 6-3. An Instrument Composed of a First-Order and a Second-Order Component

Further, we see that x' is the input to the second-order instrument. In order to determine y, we must solve the second-order model for the input

$$x = Kx_s(1 - e^{-t/\tau}); \qquad t > 0$$

PROBLEM

6-3   What is the output of a second-order instrument for an input $x(t) = Kx_s(1 - e^{-t/\tau})$?

As you can see now, it is not too difficult to obtain the response of a higher-order instrument, if we can model the instrument as a series of first- and second-order devices and if we can write an appropriate expression for the input. In the next section we will consider the frequency response of higher-order instruments, and in a later section we will see how to use this information to determine the response of an instrument to any arbitrary input.

Frequency response

Now that we have a general method of determining the response of complex instruments, we will consider the response of these instruments to a particular input—the sinusoid. As we will see in a later section of this chapter, information concerning the frequency response of an instrument can be readily used to predict its response to virtually any other input.

Let us assume that the input to the instrument shown in Fig. 6-3 is

$$x = A \sin \omega t$$

What is the output?

From Eq. (5-20) we know that x', the output of the first order component, is,

$$x' = \frac{K_1 A}{\sqrt{1 + \tau^2\omega^2}} \sin (\omega t + \phi_1)$$

$$\phi = \tan^{-1}(-\omega\tau)$$

$$(6\text{-}3)$$

Equation (6-3) represents the input to the second-order component.

The output of the second-order device is given by Eq. (5-26).

$$y = \frac{K_1 K_2 A \sin(\omega t + \phi_1 + \phi_2)}{\sqrt{1 + \tau^2 \omega^2} \sqrt{\left[1 - \left(\frac{\omega}{\omega_n}\right)^2\right]^2 + \left(\frac{2\zeta\omega}{\omega_n}\right)^2}}$$

$$\phi = \tan^{-1} \frac{(-2\zeta\omega/\omega_n)}{\left[1 - \left(\frac{\omega}{\omega_n}\right)^2\right]} \tag{6-4}$$

If we form the dimensionless ratio of output to input we obtain

$$\frac{y}{K_1 K_2 A} = \frac{\sin(\omega t + \phi_1 + \phi_2)}{\sqrt{1 + \tau^2 \omega^2} \sqrt{\left[1 - \left(\frac{\omega}{\omega_n}\right)^2\right]^2 + \left(\frac{2\zeta\omega}{\omega_n}\right)^2}} \tag{6-5}$$

We will call

$$\frac{1}{\sqrt{1 + \tau^2 \omega^2}} = \text{The amplitude ratio of the first-order component} = M_1$$

$$\frac{1}{\sqrt{\left[1 - \left(\frac{\omega}{\omega_n}\right)^2\right]^2 + \left(\frac{2\zeta\omega}{\omega_n}\right)^2}} = \text{The amplitude ratio of the second-order component} = M_2$$

Thus, Eq. (6-5) is

$$\frac{y}{K_1 K_2 A} = M_1 M_2 \sin(\omega t + \phi_1 + \phi_2)$$

We can express the output simply by stating the amplitude ratio and the phase angle

$$M \angle \phi = M_1 M_2 \angle (\phi_1 + \phi_2) \tag{6-6}$$

What can we say about the response of a composite instrument to sinusoidal inputs? Equation (6-5) shows that the output will be sinusoidal with the same frequency of oscillation as the input. This equation also indicates that the amplitude will be altered and that there will be a phase shift. Equations (6-5) and (6-6) show that the amplitude ratio of the composite instrument is equal to the product of the amplitude ratios of the components

$$M = \prod_{i=1}^{n} M_i \tag{6-7}$$

The phase angle is the sum of the individual phase angles

$$\phi = \sum_{i=1}^{n} \phi_i \tag{6-8}$$

The logarithmic plot

In many cases we find it more convenient to represent frequency response information graphically rather than numerically, as in Eq. (6-5). While we could plot on rectangular coordinates as was done in Chapter 5, another coordinate system is more convenient. Consider the instrument shown in Fig. 6-3. If we wish to show the frequency response of this device we would plot two curves. The first of these would be the amplitude ratio $(M_1 \, M_2)$ versus input frequency. The second would be phase angle $(\phi_1 + \phi_2)$ versus input frequency.

The phase angle curve is simple to construct. We can plot the phase shift for the first-order component and the shift for the second-order component individually. The phase shift for the instrument is the sum of the shifts of the components. This summation is easy to accomplish graphically; all we do is add the ordinates on the graph. You have probably done this sort of thing before.

What about the amplitude ratio curves? According to our discussion, the amplitude ratio for a complex instrument is the product of the component amplitude ratios. There does not appear to be any advantage to constructing the amplitude ratio curves graphically, since we cannot multiply with the ease that we can make a graphical summation.

Is there any transformation of coordinates, which we can make, that will help? Suppose we take the logarithm of Eq. (6-7).

$$\log M = \log \left(\prod_{i=1}^{n} M_i \right) = \log M_1 + \log M_2 + \cdots + \log M_n \tag{6-9}$$

If we plot $\log M$ versus input frequency for each component, we will be able to determine the amplitude ratio graphically by summing the ordinates on our logarithmic plot.

Although it is adequate to plot the frequency response curves as $\log M$ and ϕ versus ω, we usually use slightly different coordinates that make our curves more useful. We are usually interested in input frequencies that may vary from near zero (D.C.) to perhaps several thousand rad/sec. When a variable changes by several orders of magnitude over its range, we obtain a better representation if the logarithm of the variable is plotted. This is the case for our frequency response curves. The abscissa may be simply $\log \omega$, or we may choose some dimensionless quantity. If we choose to plot the input frequency dimensionlessly, $\log \omega\tau$ can be used for first-order instruments and $\log \omega/\omega_n$ can be used for second-order instruments. Although the dimensionless representation is often chosen, it is more convenient for us simply to use $\log \omega$.

Amplitude ratios are usually plotted in decibel units. The *decibel* is defined as twenty times the base 10 logarithm of M. Thus,

$$\textbf{db} = 20 \log M$$

You may wish to work the following two problems that will give you a convenient tool to be used in working later problems. The curves that you will plot on dimensionless coordinates are called *Bode* plots. Note that the dimensionless input quantities can be transformed to frequency by dividing by the time constant for the first-order plot or multiplying by natural frequency for the second-order plot. Your results should look something like Figs. 6-4 and 6-5.

PROBLEMS

6-4   Plot the Bode diagram for a first-order instrument. The amplitude ratio (y/KX) is to be plotted in decibels (**db**). Use the dimensionless frequency $\omega\tau$ and plot $\log \omega\tau$. This should be done for values of $\omega\tau$ ranging from 0.1–100. The phase angle is to be plotted in degrees.

6-5   Plot the Bode diagram for second-order instruments. Plot M in db versus $\log \omega/\omega_n$. Use the following values of damping ratio: 0, 0.2, 0.4, 0.6, 1.0, 1.5, and 3. Do this for values of ω/ω_n ranging from 0.1–1000.

Two other terms, the octave and the decade, are also used in conjunction with the Bode diagram. Both of these describe changes in frequency. An *octave* is defined as a range of frequencies $\omega_1 \leq \omega \leq \omega_2$

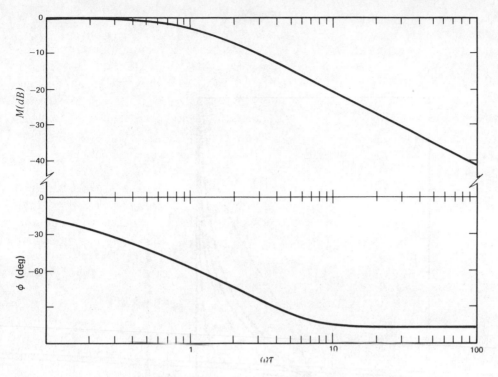

Figure 6-4. Bode Diagram for a First-Order Instrument

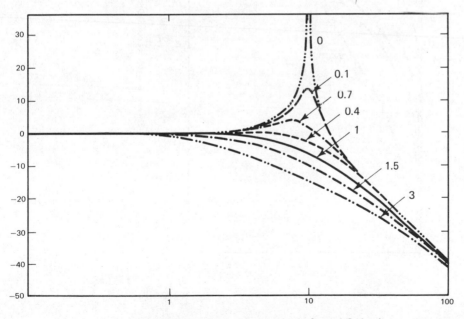

Figure 6-5(a). Bode Diagram for a Second-Order Instrument

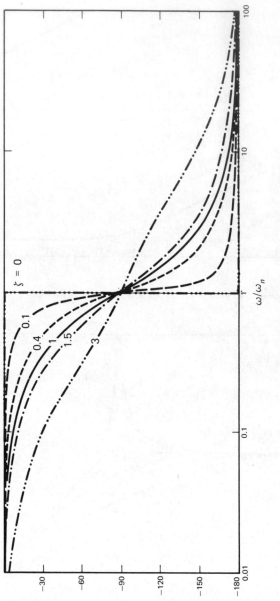

Figure 6-5(b). Second-Order Frequency Response Phase Angle

such that $\omega_2/\omega_1 = 2$. Increasing a frequency one octave means to double it. It should be obvious that the band width $(\omega_2 - \omega_1)$ of an octave is not constant; rather, it is a function of frequency. Thus, the frequency band from 1–2 rad/sec is an octave, as is the frequency band from 50–100 rad/sec.

One *decade* is the frequency band such that $\omega_2/\omega_1 = 10$. The frequency bands 1–10, 2–20, 100–1000 and 1000–10,000 rad/sec are all decades. Note once again that the width of the frequency band of one decade is not constant, but dependent upon frequency.

Since the decibel unit is probably not familiar to you, the following exercises are included. Their purpose is to give you some practice in manipulating these quantities and to give you a feel for the magnitude of change of amplitude ratio when it is expressed in decibels.

PROBLEMS

6-6 Express the following amplitude ratios in decibels:
     $M = 1.0$, $M = 1.05$, $M = 1.5$, $M = 0.95$, $M = 0.90$, $M = 0.50$.

6-7 An amplitude ratio that is initially unity decreases by 3 **db**. What is the new value of amplitude ratio?

6-8 The following are amplitude ratios expressed in **db**. Convert these to ratios. 0 **db**, 8 **db**, -6 **db**, 20 **db**, -40 **db**, -60 **db**.

The following example illustrates the use of the logarithmic plots to obtain the frequency response of higher-order instruments.

An instrument is composed of a first-order sensing element and a second-order data presentation device. The time constant for the first-order element is 0.01 sec and the static sensitivity is 2 mV per degree. The second order device has a natural frequency, $\omega_n = 120$ rad/sec, damping ratio, $\zeta = 0.4$ and the static sensitivity is 10 mm/mV. The sinusoidal transfer functions for this instrument are shown in Fig. 6-6.

We see from the figure that $M_1 = 1/[1 + (.01\omega)^2]^{1/2}$ and

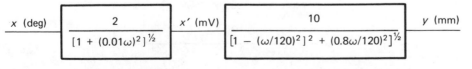

Figure 6-6. Transfer Function for a Composite Instrument

$$M_2 = \frac{1}{\left[\left(1 - \left(\frac{\omega}{120}\right)^2\right)^2 + \left(\frac{.8\omega}{120}\right)^2\right]^{1/2}}.$$

The phase angles are $\phi_1 = \tan^{-1}(-.01\omega)$ and

$$\phi_2 = \tan^{-1}\frac{\left(-\frac{.8\omega}{120}\right)}{\left[1 - \left(\frac{\omega}{120}\right)^2\right]}.$$

Our first step is to plot the amplitude ratios (in **db**) and phase angles (in degrees) versus log ω. These plots are shown in Fig. 6-7 and Fig. 6-8. Next, we sum the ordinates to obtain the amplitude ratio and phase angle for the combined components. This, remember, can be done with dividers. The resulting curves are shown in Figs. 6-9 and 6-10.

Suppose we wish to know the amplitude ratio in output units/input unit (mm/° in this case)? We know that

$$M = M_1 M_2 = \frac{y}{K_1 K_2 x}$$

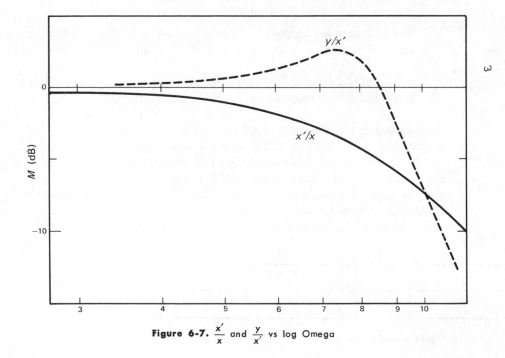

Figure 6-7. $\dfrac{x'}{x}$ and $\dfrac{y}{x'}$ vs log Omega

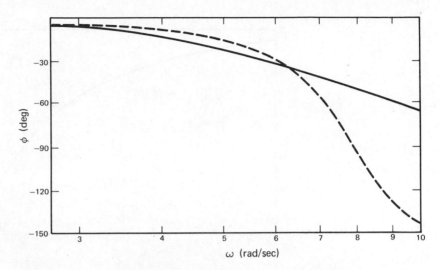

Figure 6-8. ϕ_1 and ϕ_2 vs Omega

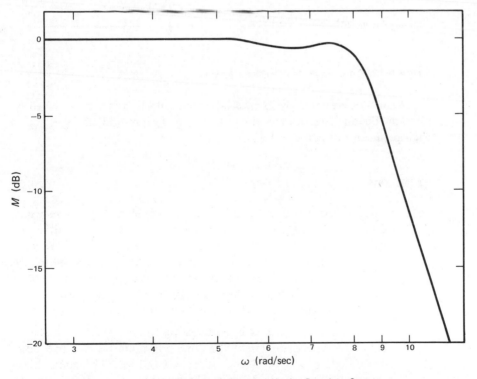

Figure 6-9. Bode Diagram for the Complete System

145

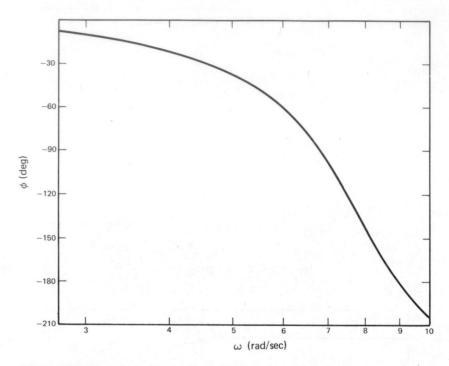

Figure 6-10. Phase Angle for Combined System

To obtain y/x, we multiply M by $K_1 K_2$. Multiplication is accomplished by addition on our logarithmic plots. Thus, we simply add $20 \log K_1 K_2$ to each point on our curve in Fig. 6-9.

PROBLEMS

6-9   A system for measuring pressure is composed of a pressure transducer ($K = .2$ mV/psi, $\omega_n = 3,000$ Hz, $\zeta = 0.1$) and an oscillograph type of recorder ($K = .25$ in./mV, $\omega_n = 3500$ Hz, $\zeta = 0.65$). Plot the frequency response for this combination. If the input amplitude is 20 psi, what is the displacement of the trace on the chart when $\omega = 1000$ Hz? when $\omega = 2000$ Hz? If the amplitude ratio is to be 1.00 ± 0.05, what is the maximum allowable input frequency?

6-10   A strip-chart recorder is considered to behave as two first-order devices connected in series. The time constants for the two components are $\tau_1 = 0.0025$ sec and $\tau_2 = 0.001$ sec. The static sensitivity of the recorder is 2 cm/V. This recorder is used to record the output of an accelerometer ($K = 0.1$ V/g, $\omega_n = 700$ rad/sec, $\zeta = 0.01$). Plot the frequency response of this instrument. What is the maximum value of input frequency if it is required that $M = 1.00 \pm 0.02$?

**6-4
Response to
Other Inputs**

You may have wondered why we have placed so much emphasis on the frequency response of composite instruments. The remainder of this chapter is intended to show you how we can use frequency response information to predict the response of a composite instrument to any input. We will begin by describing the various types of input signals that might occur and then we will see how each can be expressed in terms of frequency content.

Periodic, transient, and random signals

What kind of inputs can we expect? After some thought you will agree that we can define three categories of input that will include all possibilities. These are the periodic signal, the transient signal, and the random signal.

A *periodic* signal is a function of time that repeats itself at constant time intervals, as shown in Fig. 6-11(a). We might expect to encounter periodic signals in systems that contain parts that rotate or oscillate at constant frequency. The function $X = A \sin \omega t$ is periodic.

A *transient* signal is one that is not cyclic and becomes and remains zero after some finite period of time, as shown in Fig. 6-11(b). The pressure disturbance resulting from an explosion is an example of this kind of signal.

A *random* signal is continuous, but not cyclic. It has no particular amplitude and no specific frequency or period. If we were to observe a random signal over a long period of time, we would see that it does not repeat itself. Figure 6-11(c) is a random signal generated by a random noise generator.

Each of these signals can be transformed in such a fashion that they are represented in terms of sinusoids of varying frequency. We shall see how to perform these transformations in the following sections.

Of what use is this frequency information? If we know the input signal in terms of the frequencies it contains and the frequency response of the instrument, we can determine the output in terms of the frequencies it contains. If we can, by any means, perform an appropriate

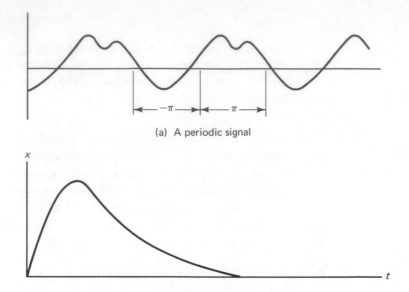

(a) A periodic signal

(b) A transient signal

(c) A random signal

Figure 6-11. Input Signals

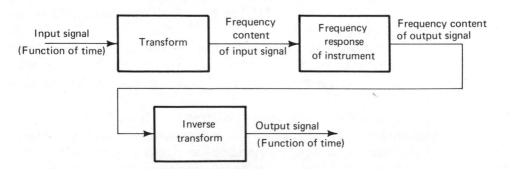
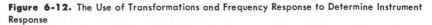

Figure 6-12. The Use of Transformations and Frequency Response to Determine Instrument Response

inverse transformation upon this output, we have the response of the instrument to the input in question. This is shown schematically in Fig. 6-12. I shall describe the details of the procedure for each type of input in the following sections.

Treatment of periodic signals

A periodic signal is shown in Fig. 6-11(a). The *period* of such a signal is the length of time required to complete one cycle of the signal. The *fundamental frequency* is the reciprocal of the period.

Let us designate the period as T. We can then state that

$$x(t) = x(t + T)$$

This signal can be represented as a series

$$x(t) = \frac{1}{2}a_o + \sum_{n=1}^{\infty} \left(a_n \cos \frac{2\pi}{T}nt + b_n \sin \frac{2\pi}{T}nt \right) \qquad (6\text{-}10)$$

We call this series the *Fourier series representation* of $x(t)$; $x(t)$ must satisfy two conditions for Eq. (6-10) to be valid. The integral $\int_{-\pi}^{\pi} |x(t)| \, dt$ must be finite and $x(t)$ must have a finite number of discontinuities. These are called the *Dirichlet conditions*. Fortunately, almost all signals of interest satisfy these conditions.

Expressions for the coefficients of Eq. (6-10) are easily derived. I will simply present the results here as the derivations appear in many references.

$$a_o = \frac{2}{T} \int_{-T/2}^{T/2} x(t) \, dt \qquad (6\text{-}11)$$

$$a_n = \frac{2}{T} \int_{-T/2}^{T/2} x(t) \cos \frac{2\pi}{T} nt \, dt \qquad (6\text{-}12)$$

$$b_n = \frac{2}{T} \int_{-T/2}^{T/2} x(t) \sin \frac{2\pi}{T} nt \, dt \qquad (6\text{-}13)$$

Since there is no necessity for the interval of integration to be symmetrical about the origin, you will find the ways in which Eqs. (6-11)–(6-13) are expressed may vary. These other expressions are valid as long as the integration is taken over one period.

We now have an exact representation of the periodic function $x(t)$. It is not convenient (or even possible) to write an infinite number of terms so we must approximate $x(t)$ by using a finite number of terms of Eq. (6-10). We cannot give the optimum number of terms to carry. We have to decide based upon the necessary accuracy and the effort involved in generating and manipulating additional terms in the series.

Now you might ask, just how can we obtain the approximate Fourier series for a given periodic function? If we know $x(t)$ explicitly, we can integrate Eqs. (6-11), (6-12), and (6-13) to determine the coefficients. How do we proceed if $x(t)$ is simply a record of the output of some instrument? A number of procedures have been worked out for evaluating the coefficients. They all require that $x(t)$ be evaluated at a discrete number of points. It is then possible to integrate Eqs. (6-11), (6-12), and (6-13) numerically. Other procedures (the methods of 12 and 24 ordinates, for instance) are described in several numerical analysis books.

Let us write the Fourier series for a square wave, as shown in Fig. 6-13. This function can be expressed

$$x(t) = -A \qquad -T/2 < t < 0$$
$$x(t) = A \qquad 0 < t < T/2$$

From Eq. (6-11)

$$a_o = \frac{2}{T} \int_{-T/2}^{T/2} x(t)\, dt$$

$$= \frac{2}{T} \int_{-T/2}^{0} -A\,dt + \frac{2}{T} \int_{0}^{T/2} A\,dt = 0$$

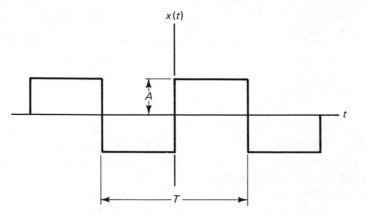

Figure 6-13. A Square Wave

The coefficient a_o is the mean value of $x(t)$. You can see from Fig. 6-13 that the mean value of the square wave is indeed zero. If we translate the function upward A units so that

$$x(t) = 0 \qquad -T/2 < t < 0$$
$$x(t) = 2A \qquad 0 < t < T/2$$

the average value of the function is A. You should verify this by integrating Eq. (6-11).

Equation (6-12) gives the values of the a_n coefficients

$$a_n = \frac{2}{T} \int_{-T/2}^{0} -A \cos \frac{2\pi}{T} nt \, dt + \frac{2}{T} \int_{0}^{T/2} A \cos \frac{2\pi}{T} nt \, dt$$

$$= \frac{A}{n\pi} \sin \frac{2\pi}{T} nt \Big|_{-T/2}^{0} + \frac{A}{n\pi} \sin \frac{2\pi}{T} nt \Big|_{0}^{T/2}$$

$$a_n = 0 - \left(-\frac{A}{n\pi}\right) \sin \frac{2\pi n}{T}(-T/2)$$

$$+ \frac{A}{n\pi} \sin \frac{2\pi n}{T}(T/2) - 0$$

$$= \frac{A}{n\pi} \sin (n\pi) + \frac{A}{n\pi} \sin (-n\pi)$$

Noting that

$$\sin n\pi = \sin (-n\pi) = 0$$

for integer values of n, we see that

$$a_n = 0$$

Integration of Eq. (6-13) gives us the b_n coefficients

$$b_n = \frac{2}{T} \int_{-T/2}^{0} -A \sin \frac{2\pi}{T} nt \, dt + \frac{2}{T} \int_{0}^{T/2} A \sin \frac{2\pi}{T} nt \, dt$$

$$b_n = \frac{A}{n\pi} \cos \frac{2\pi nt}{T} \Big|_{-T/2}^{0} - \frac{A}{n\pi} \cos \frac{2\pi nt}{T} \Big|_{0}^{T/2}$$

$$= \frac{A}{n\pi}[1 - \cos (-n\pi) - \cos (n\pi) + 1]$$

$$b_n = \frac{A}{n\pi}[2 - \cos (n\pi) - \cos (-n\pi)]$$

For odd values of n

$$\cos (n\pi) = \cos (-n\pi) = -1$$

$$b_n = \frac{4A}{n\pi} \qquad n = 1, 3, 5, \ldots \qquad (6\text{-}14)$$

For even values of n

$$\cos (n\pi) = \cos (-n\pi) = 1$$

$$b_n = 0 \qquad n = 2, 4, 6, \ldots$$

When we substitute our result for the Fourier coefficients into Eq. (6-10) we find that the Fourier representation of the square wave is

$$x(t) = \sum \frac{4A}{n\pi} \sin \frac{2\pi}{T} nt \qquad n = 1, 3, 5, \ldots, \infty$$

We call the frequency of the square wave $(2\pi/T)$ the fundamental frequency. The multiples of the fundamental frequency are called the *harmonics*. The square wave contains the odd harmonics only.

Notice that we could have saved ourselves some work if we had been able to foresee that the coefficients of the cosine terms of Eq. (6-10) were all zero. Fortunately, we can determine this in advance if we are able to identify the function as odd or even. A function is said to be *odd* if $x(-t) = -x(t)$. For *even functions*, $x(-t) = x(t)$. The Fourier series for an odd function contains only sine terms (the a_n coefficients are zero). Only cosine terms appear in the series if the function is even (the b_n coefficients are zero). Look at the square wave shown in Fig. 6-13. Here we have $x(-t) = -x(t)$, so the function is odd and every a_n is zero.

To gain some idea of the error involved in using a series with a finite number of terms, work the following problem.

PROBLEM

6-11   A square wave with $A = 1$ and $T = 1$ sec is approximated by (a) a three term Fourier series and (b) a five term Fourier series. Plot the square wave and the two approximate functions.

We do not often know $x(t)$ explicitly. More likely we have only the output of some recording device. The ideas that we have outlined in the previous paragraphs still apply, except that numerical or graphical techniques must be applied. Procedures for numerical integration are not difficult. Instead of presenting them here, any of the many texts on Numerical Analysis is recommended for reference.

Rather than integrating Eqs. (6-11)–(6-13) to obtain the Fourier coefficients, we could compute a finite number of coefficients such that the approximate Fourier series is exactly equal to $x(t)$ at a finite number of points. There will be some error at all other values of time. The magnitude of this error depends upon the number of points selected and their location. This procedure is described in the following example.

Let us write the Fourier series for the square wave shown in Fig. 6-13 if we require that the error (difference between the square wave and the value of $x(t)$ given by Eq. (6-10)) be zero at $t = T/12, T/4, 5T/12, T/2, 7T/12, 3T/4, 11T/12$. At each of these values of t we can write Eq. (6-10), thus yielding seven independent equations. If the number of coefficients to be determined is exactly seven, we can solve our set of equations for them. For the values of time given, the equations are

$$x(T/12) = \frac{1}{2}a_0 + a_1 \cos \frac{\pi}{6} + a_2 \cos \frac{\pi}{3} + a_3 \cos \frac{\pi}{2}$$

$$+ b_1 \sin \frac{\pi}{6} + b_2 \sin \frac{\pi}{3} + b_3 \sin \frac{\pi}{2} = A$$

$$x(T/4) = \frac{1}{2}a_0 + a_1 \cos \frac{\pi}{2} + a_2 \cos \pi + a_3 \cos \frac{3\pi}{2}$$

$$+ b_1 \sin \frac{\pi}{2} + b_2 \sin \pi + b_3 \sin \frac{3\pi}{2} = A$$

$$x(5T/12) = \frac{1}{2}a_0 + a_1 \cos \frac{5\pi}{6} + a_2 \cos \frac{5\pi}{3} + a_3 \cos \frac{5\pi}{2}$$

$$+ b_1 \sin \frac{5\pi}{6} + b_2 \sin \frac{5\pi}{3} + b_3 \sin \frac{5\pi}{2} = A$$

$$x(T/2) = \frac{1}{2}a_0 + a_1 \cos \pi + a_2 \cos 2\pi + a_3 \cos 3\pi$$

$$+ b_1 \sin \pi + b_2 \sin 2\pi + b_3 \cos 3\pi = 0$$

$$x(7T/12) = \frac{1}{2}a_0 + a_1 \cos \frac{7\pi}{6} + a_2 \cos \frac{7\pi}{3} + a_3 \cos \frac{7\pi}{2}$$

$$+ b_1 \sin \frac{7\pi}{6} + b_2 \sin \frac{7\pi}{3} + b_3 \cos \frac{7\pi}{2} = -A$$

$$x(3T/4) = \frac{1}{2}a_0 + a_1 \cos \frac{3\pi}{2} + a_2 \cos 3\pi + a_3 \cos \frac{9\pi}{2}$$

$$+ b_1 \sin \frac{3\pi}{2} + b_2 \sin 3\pi + b_3 \sin \frac{9\pi}{2} = -A$$

$$x(11T/12) = \frac{1}{2}a_0 + a_1 \cos\frac{11\pi}{6} + a_2 \cos\frac{11\pi}{3} + a_3 \cos\frac{11\pi}{2}$$
$$+ b_1 \sin\frac{11\pi}{6} + b_2 \sin\frac{11\pi}{3} + b_3 \sin\frac{11\pi}{2} = -A$$

This set of equations can be solved by any one of several methods. The resulting solutions are

$$
\begin{array}{ll}
a_0 = 0 & b_1 = 1.333 \ A \\
a_1 = 0 & b_2 = 0 \\
a_2 = 0 & b_3 = 0.333 \ A \\
a_3 = 0 &
\end{array}
$$

The Fourier series that approximates the square wave is

$$x(t) = 1.333 \ A \sin\frac{2\pi}{T}t + 0.333 \ A \sin\frac{6\pi}{T}t$$

If we calculate b_1 and b_3 from Eq. (6-13), the approximate series is

$$x(t) = 1.237 \ A \sin\frac{2\pi}{T}t + 0.425 \ A \sin\frac{6\pi}{T}t$$

These two equations are plotted in Fig. 6-14. Also shown is the square wave. The approximation is rough since we have carried only two terms in the series.

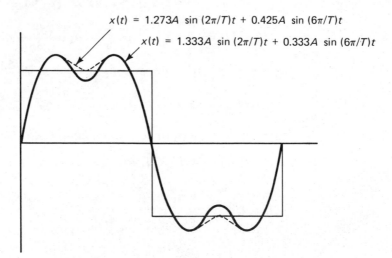

$x(t) = 1.273A \ \sin (2\pi/T)t + 0.425A \ \sin (6\pi/T)t$

$x(t) = 1.333A \ \sin (2\pi/T)t + 0.333A \ \sin (6\pi/T)t$

Figure 6-14. A Square Wave and Two Fourier Approximations

PROBLEMS

6-12 Write the Fourier series approximations of the periodic signals shown in Fig. P6-12. Your instructor will specify the number of terms to be computed and the method of computation.

6-13 The following table gives the ordinates of a periodic pressure signal at various times through one period. Write the Fourier approximation of the function using the number of terms and method specified by your instructor.

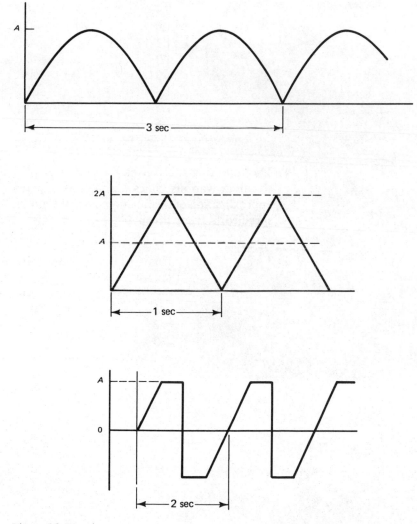

Figure P6-12.

Time (sec)	Pressure (psi)
0.1	0
0.2	20
0.3	72
0.4	93
0.5	90
0.6	78
0.7	78
0.8	71
0.9	41
1.0	40
1.1	0
1.2	−48
1.3	−60
1.4	−61
1.5	−47
1.6	0

6-14 The following table gives the ordinates of a periodic temperature signal at various times through one period. Write the Fourier approximation of the function using the number of terms and method specified by your instructor.

Time (hours)	Temperature (deg F)
1	80
2	155
3	258
4	289
5	300
6	295
7	290
8	282
9	280
10	225
11	170
12	121
13	90
14	80

Now that we have expressed the input signal as a summation of sine and cosine waves of varying amplitude and frequency, how can we use that information and the instrument frequency response to determine the output signal? Let us consider a simple example to see how this might be accomplished.

The approximate Fourier series for the function shown in Fig. 6-15 is

$$x(t) = \sin t + 0.5 \sin 3t + 0.25 \sin 5t \qquad (6\text{-}15)$$

What is the output of a second-order instrument with $\omega_n = 2$ rad/sec, $\zeta = 0.4$ and $K = 1$ if the input is that shown in Fig. 6-15.

Using Eq. (5-28) we can compute the portion of the output that is due to each input term. The output $y(t)$ is the sum of these individual output terms. That is,

$$y(t) = \frac{KX_1}{\sqrt{\left[1 - \left(\frac{\omega_1}{\omega_n}\right)^2\right]^2 + \left(\frac{2\zeta\omega_1}{\omega_n}\right)^2}} \sin(\omega_1 t + \phi_1)$$

$$+ \frac{KX_2}{\sqrt{\left[1 - \left(\frac{\omega_2}{\omega_n}\right)^2\right]^2 + \left(\frac{2\zeta\omega_2}{\omega_n}\right)^2}} \sin(\omega_2 t + \phi_2)$$

$$+ \frac{KX_3}{\sqrt{\left[1 - \left(\frac{\omega_3}{\omega_n}\right)^2\right]^2 + \left(\frac{2\zeta\omega_3}{\omega_n}\right)^2}} \sin(\omega_3 t + \phi_3)$$

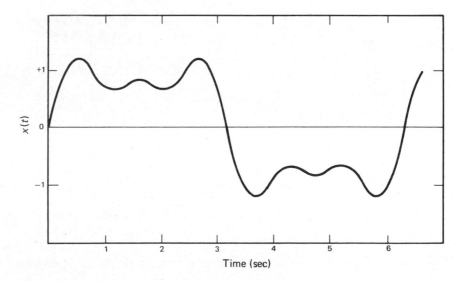

Figure 6-15. The Input $x = \sin t + .5 \sin 3t + .25 \sin 5t$

The various values of X and ω can be determined from Eq. (6-15) and are

$$\omega_1 = 1 \text{ rad/sec} \qquad X_1 = 1$$
$$\omega_2 = 3 \text{ rad/sec} \qquad X_2 = 0.5$$
$$\omega_3 = 5 \text{ rad/sec} \qquad X_3 = 0.25$$

The phase angles are given by

$$\phi = \tan^{-1} \frac{-2\zeta \dfrac{\omega}{\omega_n}}{(1 - \omega^2/\omega_n^2)}$$

K, ζ, and ω_n are properties of the instrument.

For the instrument and input we are considering, the output is

$$y(t) = \frac{1}{\sqrt{\left[1 - \left(\frac{1}{2}\right)^2\right]^2 + \left(\frac{2(0.4)1}{2}\right)^2}} \sin((1)t + \phi_1)$$

$$+ \frac{1(0.5)}{\sqrt{\left[1 - \left(\frac{3}{2}\right)^2\right]^2 + \left(\frac{2(0.4)3}{2}\right)^2}} \sin(3t + \phi_2)$$

$$+ \frac{1(0.25)}{\sqrt{\left[1 - \left(\frac{5}{2}\right)^2\right]^2 + \left(\frac{2(0.4)5}{2}\right)^2}} \sin(5t + \phi_3)$$

where

$$\phi_1 = \tan^{-1} \frac{-2(0.4)(1/2)}{1 - \left(\frac{1}{2}\right)^2} = -.49 \text{ rad}$$

$$\phi_2 = \tan^{-1} \frac{-2(0.4)(3/2)}{1 - \left(\frac{3}{2}\right)^2} = -1.43 \text{ rad}$$

$$\phi_3 = \tan^{-1} \frac{-2(0.4)(5/2)}{1 - \left(\frac{5}{2}\right)^2} = -2.78 \text{ rad}$$

The above equations reduce to

$$y(t) = 1.1765 \sin(t - 0.49) + 0.2886 \sin(3t - 1.43)$$
$$+ .0445 \sin(5t - 2.78) \tag{6-16}$$

Note that each of the terms in Eq. (6-16) corresponds to one of the terms

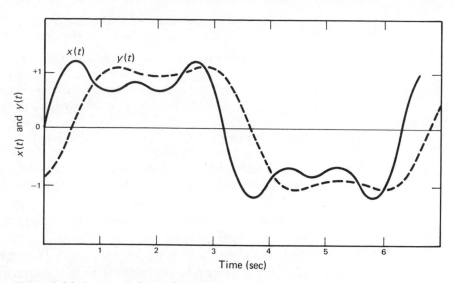

Figure 6-16. Input and Output Signals

in Eq. (6-15). The input and output of the instrument are plotted as functions of time in Fig. 6-16.

The procedure for predicting the output of an instrument when a periodic input is imposed is not too difficult. We first determine the approximating Fourier series. Then, using the frequency response of the instrument, we determine its response to each term of the series. The output of the instrument is the sum of these individual responses.

PROBLEMS

6-15 The functions described in Prob. 6-12 are each imposed on a first-order instrument with time constant of 0.8 sec and static sensitivity of 8 mV/in. In each case, A = 20 mV. What is the output of this instrument?

6-16 The functions described in Prob. 6-12 are each imposed upon a second-order instrument, which has the following dynamic characteristics:

$$\omega_n = 5 \text{ rad/sec}; \ \zeta = 0.3, \ K = 1 \text{ V/in.}$$

What is the output of the instrument?

6-17 The function described in Prob. 6-13 is imposed on a composite instrument composed of a first-order component with $\tau = 5$ sec,

$K = 10$ mV/psi and a second-order component with $\omega_n = 100$ Hz, $\zeta = 1.5$, $K = 4$ in/Volt. What is the output of this instrument?

6-18   A liquid-in-glass thermometer with a time constant of 2 minutes is used to measure the temperature signal described in Prob. 6-14. What will the maximum error (due to dynamic behavior) of the thermometer be? (Hint: Plot one cycle of the input and output.)

Transient signals

A transient signal is shown in Fig. 6-11(b). Note that the signal becomes and remains zero after some finite time. Although we cannot apply Eq. (6-10) to transient signals, we can depart from it to develop a useful method. We will not derive, but simply present the necessary equation. You can find the derivation in a number of places. If the transient signal (which we will designate $F(t)$) satisfies the following conditions, it can be represented as a function of frequency rather than as a function of time by means of the exponential Fourier transform. The conditions that must be satisfied are that:

1. $F(t)$ be defined on a specified unbounded interval.
2. $F(t)$ be sectionally continuous.
3. $F(t)$ be defined at each point of discontinuity as its mean value.
4. The integral of $|F(t)|$ must exist over the interval.

For functions that satisfy these conditions, the exponential Fourier transform is given by

$$E(\omega)\{F(t)\} = \int_{-\infty}^{\infty} F(t)e^{-i\omega t}\, dt \qquad (-\infty < \omega < \infty) \qquad (6\text{-}17)$$

Although it is necessary to consider negative values of frequency in some applications, we need not be concerned with them. We will also define the origin of the time axis so that $F(t) = 0$ for $t < 0$. These considerations allow us to write

$$E(\omega)\{F(t)\} = \int_{0}^{\infty} F(t)e^{-i\omega t}dt \qquad (0 \le \omega < \infty) \qquad (6\text{-}18)$$

Since $e^{-i\omega t} = \cos \omega t - i \sin \omega t$, Eq. (6-18) can be written

$$E(\omega)\{F(t)\} = \int_{0}^{\infty} F(t) \cos \omega t\, dt - i \int_{0}^{\infty} F(t) \sin \omega t\, dt \qquad (6\text{-}19)$$

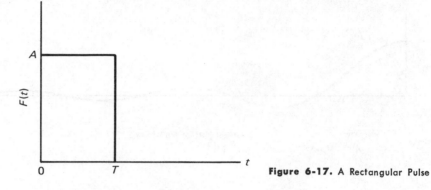

Figure 6-17. A Rectangular Pulse

The result of transforming the transient $F(t)$ is $f_e(\omega)$ which is a continuous function of frequency. Let us consider a few examples.

Figure 6-17 shows a rectangular pulse. This function can be expressed

$$F(t) = A \qquad 0 < t < T$$
$$F(t) = 0 \qquad t > 0$$
$$F(t) = A/2 \qquad t = 0 \text{ and } T$$

From Eq. (6-18)

$$f_e(\omega) = \int_0^\infty F(t)e^{-i\omega t}\, dt = \int_0^T A e^{-i\omega t}\, dt + \int_T^\infty (0)e^{-i\omega t}\, dt$$

This integration is easily accomplished.

$$f_e(\omega) = -\frac{A}{i\omega}e^{-i\omega t}\Big|_0^T A e^{-i\omega T} + \frac{A}{i\omega}$$

Since $e^{-i\omega T} = \cos \omega T - i \sin \omega T$ and $\dfrac{1}{i} = -i$, we can write

$$f_e(\omega) = i\frac{A}{\omega}(\cos \omega T - i \sin \omega T) - i\frac{A}{\omega}$$

$$= \frac{A}{\omega}\sin \omega T + i\frac{A}{\omega}(\cos \omega T - 1) \qquad (6\text{-}20)$$

This result is plotted in Fig. 6-18.

Note in the above example that the transform has real and imaginary

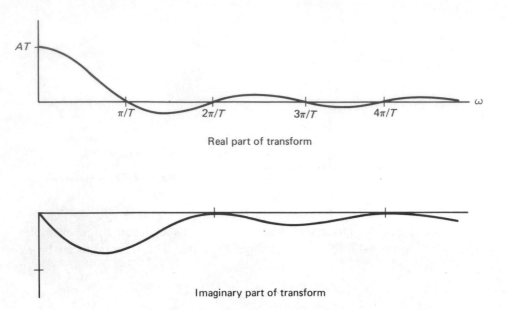

Real part of transform

Imaginary part of transform

Figure 6-18. The Fourier Transform of a Pulse

parts. It is often more informative to express $f_e(\omega)$ in terms of amplitude and phase angle. This is easy to do. The components of Eq. (6-20) are plotted in Fig. 6-19. In this figure

$$R(\omega) = \frac{A}{\omega} \sin \omega T \quad \text{(The real part of the transform)}$$

$$I(\omega) = \frac{A}{\omega} (\cos \omega T - 1) \quad \text{(The imaginary part of the transform)}$$

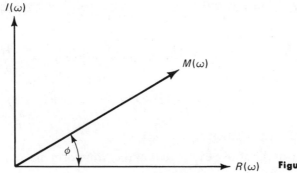

Figure 6-19. Components of $E(\omega)\{F(t)\}$

$$M(\omega) = [R(\omega)^2 + I(\omega)^2]^{1/2}$$

$$\phi = \tan^{-1}\left(\frac{I(\omega)}{R(\omega)}\right)$$

Using $M(\omega)$ and ϕ as defined in Fig. 6-18,

$$f_e(\omega) = \frac{A}{\omega}[(\sin \omega T)^2 + (\cos \omega T - 1)^2]^{1/2} \qquad (6\text{-}21a)$$

$$= \frac{A}{\omega}[2 - 2\cos \omega T]^{1/2}$$

$$\phi = \tan^{-1}\left(\frac{\cos \omega T - 1}{\sin \omega T}\right) \qquad (6\text{-}21b)$$

Figure 6-20 is a plot of the above equations. We now have the frequency content, or frequency spectrum, of the rectangular pulse. It might be helpful to think of Eqs. (6-20) and (6-21) as instruction for synthesizing the rectangular pulse. The first equation gives the frequencies present and their amplitudes. The second equation gives the phase angle at each frequency with respect to some reference.

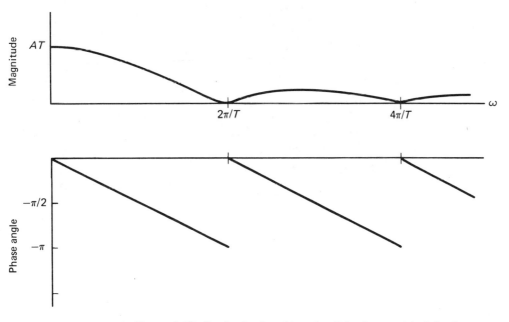

Figure 6-20. The Fourier Transform of a Pulse Expressed in Polar Form

PROBLEMS

6-19 Plot the frequency spectrum for each of the functions shown in Fig. P6-19.

6-20 Plot the frequency spectrum for each of the functions shown in Fig. P6-20.

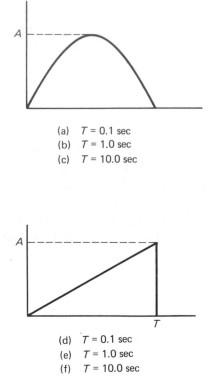

(a) $T = 0.1$ sec
(b) $T = 1.0$ sec
(c) $T = 10.0$ sec

(d) $T = 0.1$ sec
(e) $T = 1.0$ sec
(f) $T = 10.0$ sec

Figure P6-19.

In many cases, we are satisfied with knowing the frequency content of the transient. However, to be complete, let us consider how we can obtain the output of an instrument for a given input transient. Equation (6-19) gives the input to the instrument as a function of frequency. If this is multiplied by the frequency response of the instrument, the result is the output of the instrument as a function of frequency. If, for example, our rectangular pulse is the input to a first-order instrument, the output is the product of Eq. (6-21) and Eq. (5-20).

$$y = \frac{K\dfrac{A}{\omega}}{\sqrt{\tau^2\omega^2}}(2 - 2\cos \omega T)^{1/2} \angle \phi_1 + \phi_2 \tag{6-22}$$

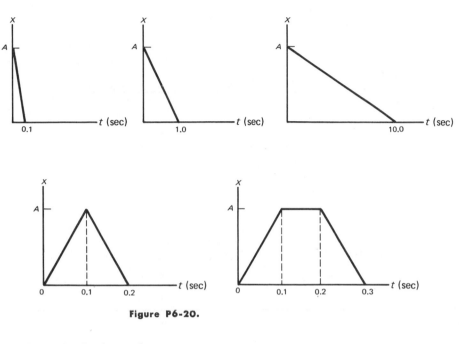

Figure P6-20.

where

$$\phi_1 = \tan^{-1}(-\omega t)$$

$$\phi_2 = \tan^{-1}\left(\frac{\cos \omega T - 1}{\sin \omega T}\right)$$

Figure 6-21 is the output of a first-order instrument for a rectangular-pulse input. Note the influence of time constant on the magnitude and phase angle of the transformed output.

If we wish to construct the instrument's output as a function of time, we must perform the inverse transform. This is defined as

$$F(t) = \frac{1}{2\pi} \lim_{\beta \to \infty} \int_0^\beta e^{i\omega t} f_e(\omega) \, d\omega \qquad (6\text{-}23)$$

$f_e(\omega)$ is the output of the instrument in terms of frequency. When you consider that this function is something like Eq. (6-22), it is obvious that Eq. (6-23) may be rather difficult to integrate. Fortunately, several algorithms are available that allow the Fourier transformation and its inverse to be accomplished on the digital computer.

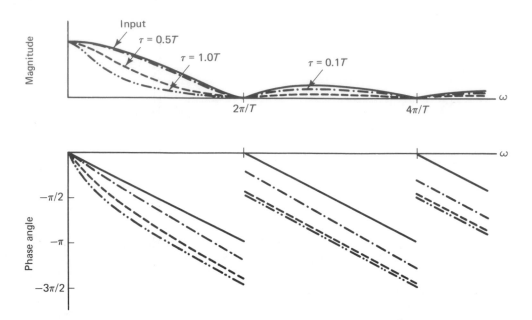

Figure 6-21. Transformed Response of a First-Order Instrument

We often speak of *fast transients* and *slow transients*. The distinction is obvious. Fast transients are of relatively short duration and slow transients are of longer duration. What statement can we make concerning the frequency content of fast and slow transient signals? If you have worked Problems 6-19 and 6-20 you should be able to answer easily. Notice that in both problems, the signals of short duration contain high-frequency components with significant amplitudes. The longer duration signals do not contain such components.

What does this mean in terms of the frequency response required of an instrument that is used to measure these transients? The fast transient requires an instrument with better (faster) dynamic response than does the slow transient. In general, we can state that as the duration of the transient decreases, our difficulty in obtaining accurate measurements increases.

Treatment of random signals

The random signal is more difficult to deal with than are periodic or transient signals. It is continuous, but it does not have a finite period. Thus, we cannot, as the result of observing the signal for some finite length of time, predict the values the signal will assume at later times.

Unlike the transient signal, which becomes and remains zero after some measureable duration, the random signal assumes (unpredictable) non-zero values for an indefinite period of time.

How can we manage such an intractible signal? Since we cannot express it as a function of time and predict its value, we must be satisfied with describing it in terms of its statistics. Of the statistics we might use, the mean value, the mean square, auto correlation function, and spectral density are the most useful.

To compute any of the above quantities, we must sample (observe) the signal for some period of time. This fact presents us with some difficulties. When should we sample? Does the random signal possess different statistics over different periods of time? (You can imagine the difficulties we might encounter if this is the case.) How many times should we sample the signal? You probably recall from Chapter 2 that the estimate of mean value and standard deviation improves as the sample size increases. This is also true for some random signals. Since we want to keep our discussion as simple as possible, we will assume that the random signals that are of interest are *stationary* (the statistics do not change with time) and *ergodic* (that is, one sample will be sufficient to determine the required statistics).

The *mean*, or *average* value, of a random signal is defined as

$$\bar{x}(t) = \lim_{T \to \infty} \frac{1}{2T} \int_{-T}^{T} x(t) \, dt \tag{6-24}$$

This is the average value of the amplitude of the signal. We are, of course, interested in this quantity, but a more useful statistic is the *mean square value* of the random signal. This is defined as

$$\overline{x^2}(t) = \lim_{T \to \infty} \frac{1}{2T} \int_{-T}^{T} x^2(t) \, dt \tag{6-25}$$

$\overline{x^2}(t)$ is often referred to as the *average power* of the random signal. If $\bar{x}(t)$ is zero, the mean square of the signal is a measure of the deviation of the signal from its average value. What if $\bar{x}(t)$ is not zero? In that case we compute

$$\overline{x^2}(t) = \lim_{T \to \infty} \frac{1}{2T} \int_{-T}^{T} [x(t) - \bar{x}(t)]^2 \, dt$$

which is also a measure of the deviation of the signal from its mean value.

Note that in both Eqs. (6-24) and (6-25), the signal is integrated over the entire time that it exists. Since it is not reasonable to do this in practice, what is a reasonable value for T? Unfortunately, we cannot answer that completely. We must compromise between our desire for accuracy and the necessity to obtain results reasonably quickly. One procedure for determining the optimum value for T is to sample the signal for a short period, say T_1 and evaluate the mean value, $\bar{x}_1(t)$. The signal is then sampled for a period T_2 ($T_2 > T_1$) and $\bar{x}_2(t)$ computed. This is repeated for $T_3, T_4 \ldots$, until we find that successive values of $\bar{x}(t)$ do not change significantly. When this occurs, we assume that the sampling time is long enough.

We need to establish some way in which the random signal can be expressed in terms of the amplitudes and frequencies of its sinusoidal components. Before we show this mathematically, we want to see how to do it experimentally. Suppose we feed a random signal into a device called a filter. This filter will allow a single frequency to pass through it and will block all others. The output of the filter is a sine wave that has constant frequency and variable amplitude. If we examine the amplitude carefully, we find that it varies randomly. It is possible to determine the mean square value of the amplitude by the process indicated in Eq. (6-25).

When this process is repeated many times, we can establish a curve that shows the mean square amplitude of every frequency present in the random signal. The area beneath this curve is the mean square amplitude of the random signal. Before we see how to use this information, let us look at a more formal development of this idea.

For our stationary ergodic random signal, the *autocorrelation* function is defined

$$R(\tau) = \lim_{T \to \infty} \frac{1}{2T} \int_{-T}^{T} x(t)x(t + \tau)\, dt \qquad (6\text{-}26)$$

According to Eq. (6-26), the autocorrelation function is determined by determining the value of the random signal at times t and $t + \tau$, multiplying them, and computing the average of this product for all values of time. The result is a function of the time delay, τ. This procedure can be performed by instruments that are designed for this purpose. The equipment shown in Fig. 6-22 is capable of computing and displaying the autocorrelation function. It can also perform several other types of statistical analysis.

Although the autocorrelation function has some interesting properties, the Fourier transform of the autocorrelation function is of

Figure 6-22. Instrumentation for Analyzing Random Signals *(Courtesy Hewlett-Packard Co.)*

more interest to us. According to Eq. (6-17), the Fourier transform of the autocorrelation function is

$$S(\omega) = \int_{-\infty}^{\infty} R(\tau)e^{-i\omega\tau}\,d\tau \qquad (6\text{-}27)$$

The inverse transform is

$$R(\tau) = \frac{1}{2\pi}\int_{-\infty}^{\infty} S(\omega)e^{i\omega t}\,d\omega \qquad (6\text{-}28)$$

Let us examine Eq. (6-28) first. Note that for $\tau = 0$

$$R(0) = \frac{1}{2\pi}\int_{-\infty}^{\infty} S(\omega)\,d\omega \qquad (6\text{-}29)$$

From Eq. (6-26) we see that $R(0)$ is the mean square value of our random signal. The function $S(\omega)$ as defined by Eq. (6-27) represents the amount of signal per unit frequency, and its integral is the mean square of the random signal. This is the function that was established using our filter. $S(\omega)$ is usually called the *power spectral density* of the random signal.

Figure 6-23 is the plot of a particular power spectral density (PSD). Note that the portion of the random signal that occurs between frequencies ω_a and ω_b is given by

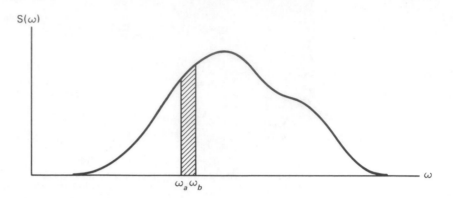

Figure 6-23. A Power Spectral Density

$$\int_{\omega_a}^{\omega_b} S(\omega)\, d\omega$$

Before we go further, let us consider what we have accomplished by our manipulations. By means of the operation defined by Eq. (6-27), we have transformed the random signal (a time function) into a function of frequency. We have accomplished this through the use of a finite sample of the random signal. Is it possible to use our knowledge of frequency response to predict the behavior of an instrument when a signal of known PSD is imposed upon it?

We will not attempt to prove the relationship we are seeking, but let us demonstrate the principle by means of a simple example and then extend the idea to random signals. Let us first consider an instrument with frequency response, $M(\omega)$. Let the input signal be

$$x(t) = A \sin \omega t$$

The mean square value of the input signal is

$$\overline{x^2(t)} = \frac{1}{T} \int_0^T A^2 \sin^2 \omega t\, dt$$

where T is the period of the input signal. Now the output signal is

$$y(t) = M(\omega) A \sin \omega t$$

and the mean square value of the output signal is

$$\overline{y^2(t)} = \frac{1}{T} \int_0^T [M(\omega)]^2 A^2 \sin^2 \omega t\, dt$$

$$= (M(\omega))^2 \left[\frac{1}{T} \int_0^T A^2 \sin^2 \omega t\, dt \right]$$

since $M(\omega)$ is not a function of time. Notice that the term in brackets is the mean square of the input signal. Thus, we have

$$\overline{y^2}(t) = M(\omega)^2 \overline{x^2}(t) \tag{6-30}$$

Is this relationship valid if $x(t)$ contains more than one frequency? To determine this, we assume

$$x(t) = A \sin \omega t + B \sin 2\omega t$$

$$\overline{x^2}(t) = \frac{1}{T} \int_0^T (A^2 \sin^2 \omega t + 2AB \sin \omega t \sin 2\omega t + B^2 \sin^2 2\omega t) dt$$

Integrating the second term gives

$$\frac{2AB}{T} \int_0^T \sin \omega t \sin 2\omega t \, dt = \frac{2AB}{T} \left[\frac{\sin \omega t}{2} - \frac{\sin 3\omega t}{6} \right]_0^T$$

since T is the period of the signal, this term is zero. Therefore, we can write

$$\overline{y^2}(t) - [M(\omega)]^2 \frac{1}{T} \int_0^T (A^2 \sin^2 \omega t + B^2 \sin^2 2\omega t) \, dt$$

$$= [M(\omega)]^2 \overline{x^2}(t)$$

Although it is perhaps not obvious, we can extend these ideas to situations in which the frequency spectrum is continuous rather than discrete. If our random signal has an input power spectral density, $S_i(\omega)$ the output PSD is

$$S_0(\omega) = [M(\omega)]^2 S_i(\omega) \tag{6-31}$$

Once again, we find that the frequency response of the instrument can be used to predict the response of the instrument when the frequency content of the input signal is known. We cannot determine the time response of the instrument because we are unable to describe the random signal as a function of time.

Perhaps this example will help you understand how we can utilize knowledge of the PSD. Suppose an accelerometer ($\omega_n = 800$ Hz, $\zeta = 0.1$) is used to sense a random acceleration. The PSD of this acceleration is shown in Fig. 6-24. What are the mean square value of the input to the instrument, the PSD of the output signal, and the mean square of the output signal?

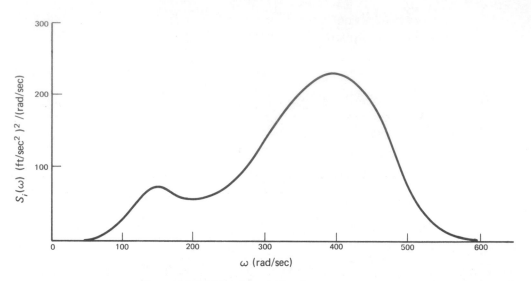

Figure 6-24. PSD of Input Signal

We can determine the mean square of the input by integrating $S_i(\omega)$. Since we do not know $S_i(\omega)$ explicitly, we must integrate graphically. This is best done by measuring the area with a planimeter and multiplying by the appropriate scale factor. The result of this integration is $\sqrt{\overline{x^2}}(t) = 5.54 \times 10^4 (\text{ft/sec}^2)^2$.

Perhaps the root mean square (rms) value of the acceleration is more meaningful: rms acceleration $= \sqrt{\overline{x^2}(t)} = 235.37$ ft/sec^2

In order to calculate the output PSD, we must multiply the square of the amplitude ratio (M^2), for the accelerometer frequency response by $S_i(\omega)$, at enough values of frequency so that the output can be plotted. Since we know the accelerometer is a second-order device, this can be accomplished easily. $S_o(\omega)$ is shown in Fig. 6-25.

PROBLEM

6-21   A random signal with the PSD shown in Fig. P6-21 is imposed on the instrument described in Prob. 6-9. Plot the output PSD and calculate the root mean square values of the input and output signals.

6-22   Would you recommend that the instrument described in Prob. 6-10 be used to measure the random acceleration that is represented by the PSD in Fig. P6-22? If not, explain and state the changes necessary to insure satisfactory performance.

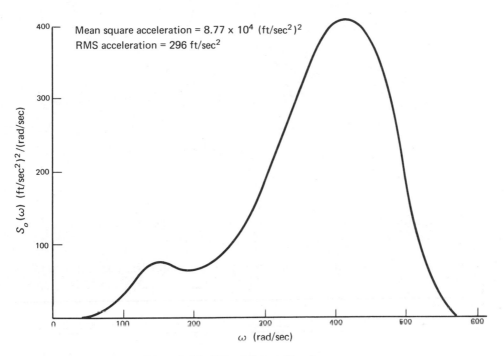

Mean square acceleration = 8.77 × 10⁴ (ft/sec²)²
RMS acceleration = 296 ft/sec²

Figure 6-25. PSD of Output Signal

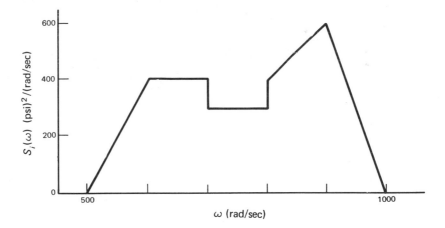

Figure P6-21.

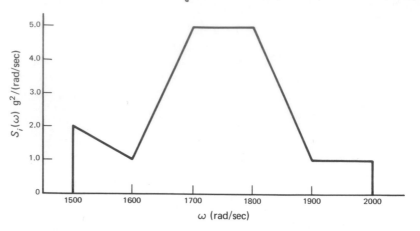

Figure P6-22.

6-5   We have covered a considerable amount of territory in this chapter.
Summary   First, we have seen that complex instruments can usually be treated
as though they are composed of a number of first- and second-order
devices connected in series. The frequency response of a complex
instrument can be determined by the use of Eqs. (6-7) and (6-8).

   In the last three sections we discovered that the three types of signals
we can expect to encounter can all be considered to be composed of
sinusoidal components. It is convenient to consider these signals in
terms of their frequency content. Since we know the frequency
response of our instrument, it is not difficult to determine the frequency content of its output. We can frequently transform the output
as a function of frequency into output as a function of time.

7

Selecting
an Instrument

You have now learned about the performance of instruments and
are aware of the nature of the input signals to expect. Let us see how
to select an instrument for a particular application. Among the factors
we should consider are the physical quantities to be measured, the
reason for measuring them, the economics involved and, of course,
the available information.

7-1
Expected Inputs

The first consideration in selecting an instrument is the expected input.
In some applications, the variable to be measured is obvious. If we
are directed to determine the speed of a conveyor belt, for example,
not much time is required to decide just what to measure. Suppose
that the task is to find the efficiency of the local power-generating
station. This presents a different problem. We have to know something
about the power station before we begin. This is not unusual. It is not
sufficient to know all about instrumentation; we have to be familiar
with the system to which the instruments will be applied.

After identifying the variables to be sensed, we must obtain the
maximum and minimum values that the inputs are expected to assume.
We will probably discover that the expected magnitude of the inputs

will eliminate certain instruments from our consideration. For example, it would be foolish to attempt to measure a temperature of 2000°F with a mercury-in-glass thermometer.

In addition to considering magnitude, we should consider the nature of the input signal. Is it constant or time dependent? If it is time dependent is it periodic, transient, or random? What is the expected frequency content? The last question cannot be answered exactly, prior to actual measurement, but we can usually make some estimate of frequency content.

For instance, suppose we are interested in measuring the vibration of an automobile engine. What kind of signal do we expect to encounter and what is its frequency content? First, the signal is most likely to be periodic since the engine is a piece of rotating machinery. The frequencies present depend upon the speed at which the engine is running. We would expect the lowest frequency present to correspond to the speed of the major rotating parts. The other frequencies we anticipate will be integral multiples of the fundamental frequency.

The reciprocating parts will also contribute substantially to the vibration. The frequency at which this occurs depends upon speed and arrangement of the crankshaft throws. For instance, if we are concerned with a 6-cylinder engine in which the pistons move in pairs, the fundamental frequency of vibration due to the reciprocating parts will be three times the frequency of rotation of the crankshaft. We will also expect harmonics of the lowest or fundamental frequency to be present.

How can we predict the frequencies that are present in a transient? Since we cannot know the exact form of the signal in advance, we cannot apply the Fourier transform. However, we can probably make a reasonable estimate of the duration of the transient. We then approximate the signal as a rectangular pulse or half sine wave of the estimated duration. The transform of our approximate signal will give us some estimate of the frequencies we should expect. Perhaps you think this is crude. If you can determine the form of the anticipated transient more accurately, you should certainly do so. In cases in which you cannot, the above procedure represents a rational way to estimate frequency content.

We have more difficulty in estimating the frequency content of random signals because of their unpredictable nature. In some cases (atmospheric turbulence, for instance) we can find descriptions of the PSD in the literature. In other cases, we can perhaps estimate the major frequencies present by careful consideration of the systems involved. Suppose we are interested in the vibrations induced in an automobile due to the roughness of the road surface. We recognize that the surface irregularities are probably random in spacing and height. If the car is traveling on a concrete pavement with uniformly spaced expansion joints, a dominant frequency in the PSD will probably be due to the joints.

7-2
Acceptable
Error

As we have seen, an instrument does not often indicate the exact value of the variable measured. The error present can stem from two situations. A portion of the error can be attributed to static inaccuracies. These static errors are described by the static characteristics of the instrument. Other errors are caused by the instrument's inability to "follow" time-dependent inputs exactly. This inability is described by the dynamic characteristics. Before we go any further let us look at a couple of examples of dynamic error.

Suppose a thermocouple (a first-order instrument) is subjected to a temperature step increase of 80°F. The time constant of the instrument is 0.1 sec. What is the dynamic error (difference between the input and output) at elapsed time of 0.01, 0.1, and 1.0 sec? Before looking at the solution, why don't you try it yourself?

From Eq. (5-17)

$$y = Kx_s(1 - e^{-t/\tau}) \tag{7-1}$$

In this case the static sensitivity K has units mV/deg since the output of a thermocouple is electrical potential. If we rewrite Eq. (7-1) as

$$\frac{y}{K} = x_s(1 - e^{-t/\tau})$$

the left side of the resultant equation is the output of the thermocouple expressed in degrees. The value of the step, x_s, is 80°. Thus, at $t = .01$ sec,

$$\frac{y}{K} = 80(1 - e^{-.01/.1}) = 80(0.095) = 7.6$$

The corresponding dynamic error is

$$\text{Error} = 7.6 - 80 = -72.4°$$

(The minus sign indicates that the output is less than the input).
In the same manner, we find that

$$Error = -25.4° \quad at \quad t = 0.1 \text{ sec}$$
$$Error = -0.074° \quad at \quad t = 1.0 \text{ sec}$$

These results are shown in Fig. 7-1.

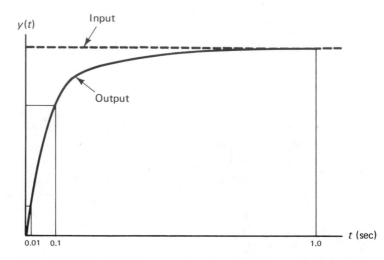

Figure 7-1. Dynamic Error

As a second example, let us consider the response of an accelerometer that has a natural frequency, $\omega_n = 10$ rad/sec and a damping ratio of 0.6, to a sinusoidal input of $40 \sin 3t$ ft/sec². If we are simply interested in the error in amplitude, we need only to consider the constant term on the right side of Eq. (5-26). Thus, the ratio of output amplitude to input amplitude is

$$\frac{y}{KX} = \frac{1}{\sqrt{\left(1 - \left(\frac{\omega}{\omega_n}\right)^2\right)^2 + \left(2\frac{\zeta\omega}{\omega_n}\right)^2}}$$

For the accelerometer we are considering

$$\frac{y}{KX} = \frac{1}{\sqrt{\left(1 - \left(\frac{3}{10}\right)^2\right)^2 + \left(\frac{2(.6)3}{10}\right)^2}} = 1.02$$

That is, if the peak input amplitude is 40 ft/sec², the peak output amplitude is 40.8 ft/sec², and the dynamic error is $40.8 - 40 = 0.8$ ft/sec².

We should note that even though the dynamic error as computed above is small (2%), the dynamic error at any instant may be much larger. Let us compute the dynamic error at $t = 0.5$, 1.0, and 2.0 sec. The output of the instrument at any time, t, is given by Eq. (5-26). Thus, for example,

$$\frac{y}{KX} = \frac{1}{\sqrt{\left(1 - \left(\frac{3}{10}\right)^2\right)^2 = \left(\frac{2(0.6)3}{10}\right)^2}} \sin\left(\omega t - .38\right)$$

If the selected values of time (0.5, 1.0, and 2.0 sec) are substituted, we find the corresponding values of y/KX. The error is again computed as the difference between the input and the output. Table 7-1 contains the results of these calculations. Note that the error at any instant of time can assume very large values even though the difference in peak input and output values is very small. This is due, of course, to the phase shift that occurs in the instrument.

TABLE 7-1

t (sec)	Input $X \sin \omega t$	Output (y/KX)	Error	% Error
0.5	0.997	0.918	—0.079	7.5
1.0	0.141	0.508	0.367	260.0
2.0	—0.279	—0.623	—0.349	125.0

The procedures we have demonstrated above are applicable to any instrument and for any input. Details of calculations may be considerably more complicated, but the error in output can be predicted for any value of time.

We are concerned with the maximum error that we can allow a particular instrument to contribute. This can be determined from a process called *error analysis*. We will not develop this topic; we will only look at a brief example.

Let us suppose that we are interested in measuring the efficiency of an engine. This involves the measurement of engine speed, torque, and the fuel consumed. Each of these measurements is subject to an error that can be predicted from the instrument characteristics. By means of the techniques of error analysis, we can determine exactly what contribution each error makes to the error on the computed value of efficiency. These contributions depend upon the functional relationship between the variables and the efficiency. If we find it necessary to reduce the error associated with the computed efficiency, we can do so by selecting instruments that exhibit smaller error. The acceptable error from each instrument can be determined from an error analysis.

7-3
Instrument
Selection

After we obtain a good estimate of the expected inputs and the allowable error, we are ready to select the instruments that are required. What do we consider? Perhaps the first condition is that the instrument be sensitive to the expected input and have the proper range. We can then check the other static characteristics. Can the instrument withstand the environment to which it will be exposed? We will usually find this stated in specifications. It is also necessary to determine whether the static sensitivity is adequate (large enough) and the resolution is small enough so that we can sense changes in input satisfactorily.

We must also consider the required accuracy. If we inspect a few catalogs and specifications, we will find that temperature sensing devices, for example, are available with accuracies ranging from ± 2.0 percent down to perhaps 0.01 percent. Which one should be selected? If an error analysis has been performed, we will have some indication. Would it be wise to select an instrument with better accuracy than is actually required? Before doing so, we should consider this. An instrument with high accuracy is usually more expensive than one with less accuracy. Usually, the more accurate instruments are more difficult to operate properly and will require the hand of skilled technicians to obtain their full potential. The task is to select an instrument that is adequate for the job at hand.

If the expected input is not constant, we must insure that the instrument's dynamic performance is adequate. The best way to do this is probably to consider the frequency response. For the best performance, we want the amplitude ratio to be constant over the expected range of input frequencies. We have learned how to predict frequency response for most instruments. We have also discussed methods of determining the frequency contents of input signals. It is not difficult to make sure that natural frequencies and time constants are properly selected. Unfortunately, many engineers fail to take this precaution. The result is often masses of useless data.

7-4
Other
Considerations

We have discussed only the most important and general factors involved in selecting instruments. There are many others. Most of these factors are associated with specific applications. Some of the references listed in Chapter 8 contain extensive discussions of this nature. The best advice for selecting instruments is to be as knowledgeable as possible about the physical systems that you are working with. Only through this means will you be certain to avoid serious errors.

Another point that we have not discussed extensively is the effect that the act of making a measurement has on a system. In Chapter 1 we noted that energy is always transferred when a physical quantity is measured. If we measure the temperature of a container of hot liquid, energy (in the form of heat) is transferred from the liquid to the thermometer. The result is that the temperature of the liquid decreases slightly. This is the temperature that we sense. The act of measurement always alters the system slightly. Our job is to see that this alteration is minimal. Once again, a thorough knowledge of the system involved is essential to accomplish this.

Often our most severe problems do not even involve the proper selection of instrumentation. These troubles are generally encountered when we have installed the instruments and are trying to make them work properly. If we have these difficulties, one can offer little advice and considerable sympathy. Usually we can rely upon our experience, the advice of experts, and the application bulletins that are mentioned in the next chapter. Good luck!

8

Where Next?

Upon reaching this point, you should be able to accomplish the goal which we set before you in the introduction:

> *Given a manufacturer's specification sheet, you will be able to describe the way that an instrument will respond to a static input and to various time-dependent inputs.*

This achievement may be sufficient for your particular needs. It is hoped, however, that this study has stimulated your interest in the field of instrumentation. It is our intention in this chapter, to give you some suggestions for the direction your further studies might take. We will also list some sources of information about instruments and measurements that we believe will be useful.

8-1
Areas of
Interest

We can identify three broad areas of instrumentation and measurement. These are: (1) precision measurement and calibration, (2) design, and (3) application. Although the lines bounding each of these areas of interest are not distinct, each area involves a different view of instrumentation.

The engineer involved in precision measurement and calibration is usually interested in very accurately measuring static values of physical parameters. He is interested in establishing, maintaining, and improving standards. Quite often he will be charged with the maintenance of the calibration facility for a laboratory or company and will supervise the calibration of the instrumentation used in that company. In large installations he may specialize in the measurement of a particular parameter, (mass or electrical resistance, for instance). More often, however, he will have broader interests and responsibilities. Figure 8-1 shows a facility that is used for the calibration of temperature-measuring devices.

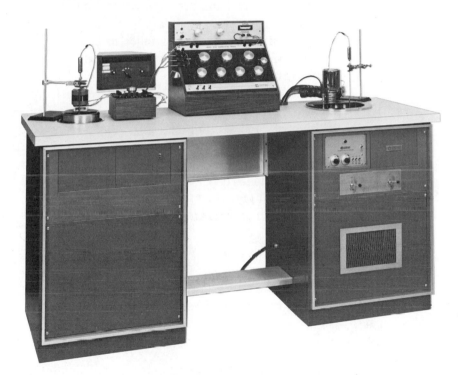

Figure 8-1. A Temperature Calibration Facility *(Courtesy Rosemount, Inc.)*

Perhaps you are interested in designing instrumentation. In this case, you will probably be more interested in the physical phenomena that can be used to sense the parameters to be measured. You will also be interested in the electronic, mechanical, and fluid systems that are often included in the instrument. Because of this diversity, you will have to become knowledgeable in many areas. In addition, you will practice the art of design in much the same way as engineers in other fields.

Most engineers who work in instrumentation are engaged in application. This might involve research in a specific area (heat transfer, for instance) or instrumenting tests for product development or to measure and certify performance. Instruments are also important components of control systems, which range from the simple thermostatic temperature control to the control of complex processing plants such as oil refineries. Quite often the engineer who is interested in some phase of application will find that he must become expert in a particular area (petroleum processing, for example) so that he can utilize this knowledge of instrumentation to the best advantage. In other cases, he must have some competence in many areas. This is most likely to occur in small organizations, which do not have a large engineering department.

Regardless of the particular area of instrumentation in which you choose to work, you will be faced with the problem of learning about the field and keeping current with developments in your area. At the present time, only a few courses in instrumentation are available at most engineering schools. These are usually quite general in nature. If you intend to learn beyond these courses, you must devise your own plan of study.

What resources are available to you? As is the case in most fields of engineering, there is a considerable fund of available literature. It is usually found in books, periodicals, and literature published by manufacturers. Let us spend some time and consider each of these sources. We wish to determine the nature of the information found in each and how it might be of use to us. Our list is not intended to be exhaustive, but should be considered as a reference guide

**8-2
Books**

There are a number of textbooks that contain general descriptions of the types of instruments that are commonly encountered. These are quite useful as they contain information about the principles of operation, some indication of the details of construction and the typical ranges and performance. Some books of this type are

1. Doebelin, E. O., *Measurement Systems: Application and Design*. New York: McGraw-Hill Book Company, 1966.

2. Beckwith, T. G. and Buck, N. L., *Mechanical Measurements*. Addison-Wesley Publishing Company, Inc., 1961.

3. Holman, J. P., *Experimental Methods for Engineers*. New York: McGraw-Hill Book Company, 1966.

4. O'Higgins, P. J., *Basic Instrumentation Industrial Measurement*. New York: McGraw-Hill Book Company, 1966.

A number of books of a more specialized nature are also available. These concern themselves with a particular measurement or class of measurements. As you would expect, they contain more detailed discussions of the instrumentation designed to sense a particular input and its application. A few books of this nature are

1. Baker, H. D., Ryder, E. A. and Baker, N. H., *Temperature Measurement in Engineering*. John Wiley and Sons, Inc., 1953.

2. Kivenson, G., *Industrial Stroboscopy*. New York: Hayden Book Company, Inc., 1965.

3. "Biomedical Sciences Instrumentation," Instrument Society of America.

4. "Marine Sciences Instrumentation," Instrument Society of America.

A third class of books deals with the design of instruments and transducers. These books contain information about the physical phenomena that can be used to sense the parameters to be measured. This type of information is of great value in devising new methods of sensing and measurement. Typical examples of these books are

1. Norton, H. N., *Handbook of Transducers for Electronic Measuring Systems*. Englewood Cliffs, N.J.: Prentice-Hall, Inc., 1969.

2. Neubert, H. K. P., *Instrument Transducers*. New York: Oxford University Press, Inc., 1963.

3. Lion, K. S., *Instrumentation in Scientific Research*. New York: McGraw-Hill Book Company, 1959.

8-3
Periodicals Although books are indispensable sources of information, they can never be completely up to date. This is because of the rather long period of time required to produce them. More current information will be found in the various periodicals that are available. There are two types of periodicals of interest: journals and other publications of various technical and professional societies and trade magazines circulated by several publishers.

Each of the technical and professional societies publishes journals and transactions containing papers and articles that are of importance and interest to its membership. Quite often, articles describing the use of instrumentation in specific applications are included. These can be very useful as the applications are generally unusual and the techniques used are unique. Unfortunately, these articles are often overlooked unless one is making a deliberate search for the literature.

Some of the societies (ASME and IEEE, for example) publish transactions that are devoted in part to instrumentation. These contain papers that have been judged to be of general interest to those working in this field. Because of the close relationship between instrumentation and automatic controls, the journals contain papers in both fields.

A few technical societies are particularly interested in instrumentation. Probably the largest of these is the Instrument Society of America (ISA). The ISA and its various divisions are responsible for a number of publications and meetings devoted to instrumentation and control. In addition, ISA develops standards and practices concerning the testing, calibration, and specification of many instruments, transducers, and components.

The ISA has a number of publications. The journal, titled *Instrumentation Technology*, contains articles of general interest. In addition, the society publishes educational materials, proceedings of its annual and various divisional meetings, and a number of divisional newsletters.

There are also a number of magazines, which are printed by private publishers, that are devoted entirely or partially to instrumentation. These usually contain articles that describe new instruments or give information concerning applications. Several of these publications are very good and are an excellent source of information. Be sure to give some attention to the advertising they contain. Often this is as useful as the editorial content of the magazine.

**8-4
Manufacturers'
Publications**

The final source of information we will discuss is the instrument manufacturer. When you consider this source, you probably think of the advertising literature and catalogs that are published. These are very valuable because they inform you of the available products, and they provide most of the data that you need concerning performance.

Manufacturer's publications are not limited to advertising and catalogs, however. Many also publish small magazines and applications bulletins. The magazines are usually short, perhaps 10 to 12 pages in length. Each issue is generally devoted to one or two instruments in the maufacturer's line. The articles will contain descriptions of the instrument and its performance. You will also find various applications of the instrument outlined in some detail.

As you might expect, the application bulletin will contain detailed information concerning some aspect of the use of a particular instrument or component. The discussion may be very specific (instructions for the use of a particular cement for bonding strain gages, for instance), or general (problems in the application of signal conditioners, for instance). In either case, the publication is generally quite useful because it addresses itself to the solution of a specific problem.

8-5
Conclusion
Our comments in this chapter have been rather general. We did not intend to compile a bibliography. Rather, we have tried to give an idea of the types of literature that are available and indicate their sources. It is up to you to locate the books, periodicals, and other publications that are of the most value to you.

Appendix

The following table gives values of the integral of Eq. (2-11) between the limits of zero and z. This integral is the area beneath the probability density function as shown in Fig. A-1. The values are computed for a distribution that has a mean value of zero and a standard deviation of 1.0. To use the table for other values of mean and standard deviation, use the transformation

$$z = \frac{x - m}{\sigma}$$

where

$x =$ value of random variable

$m =$ mean value of the distribution

$\sigma =$ standard deviation of the distribution

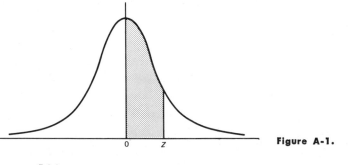

Figure A-1.

188

TABLE A-1　Normal Curve Areas

z	0	1	2	3	4	5	6	7	8	9
0.0	0.0000	0.0040	0.0080	0.0120	0.0160	0.0199	0.0239	0.0279	0.0319	0.0359
0.1	0.0398	0.0438	0.0478	0.0517	0.0557	0.0596	0.0636	0.0675	0.0714	0.0753
0.2	0.0793	0.0832	0.0871	0.0910	0.0948	0.0987	0.1026	0.1064	0.1103	0.1141
0.3	0.1179	0.1217	0.1255	0.1293	0.1331	0.1368	0.1406	0.1443	0.1480	0.1517
0.4	0.1554	0.1591	0.1628	0.1664	0.1700	0.1736	0.1772	0.1808	0.1844	0.1879
0.5	0.1915	0.1950	0.1985	0.2019	0.2054	0.2088	0.2123	0.2157	0.2190	0.2224
0.6	0.2257	0.2291	0.2324	0.2357	0.2389	0.2422	0.2454	0.2486	0.2517	0.2549
0.7	0.2580	0.2611	0.2642	0.2673	0.2704	0.2734	0.2764	0.2794	0.2823	0.2852
0.8	0.2881	0.2910	0.2939	0.2967	0.2995	0.3023	0.3051	0.3078	0.3106	0.3133
0.9	0.3159	0.3186	0.3212	0.3238	0.3264	0.3289	0.3315	0.3340	0.3365	0.3389
1.0	0.3413	0.3438	0.3461	0.3485	0.3508	0.3531	0.3554	0.3577	0.3599	0.3621
1.1	0.3643	0.3665	0.3686	0.3708	0.3729	0.3749	0.3770	0.3790	0.3810	0.3830
1.2	0.3849	0.3869	0.3888	0.3907	0.3925	0.3944	0.3962	0.3980	0.3997	0.4015
1.3	0.4032	0.4049	0.4066	0.4082	0.4099	0.4115	0.4131	0.4147	0.4162	0.4177
1.4	0.4192	0.4207	0.4222	0.4236	0.4251	0.4265	0.4279	0.4292	0.4306	0.4319
1.5	0.4332	0.4345	0.4357	0.4370	0.4382	0.4394	0.4406	0.4418	0.4429	0.4441
1.6	0.4452	0.4463	0.4474	0.4484	0.4495	0.4505	0.4515	0.4525	0.4535	0.4545
1.7	0.4554	0.4564	0.4573	0.4582	0.4591	0.4599	0.4608	0.4616	0.4625	0.4633
1.8	0.4641	0.4649	0.4656	0.4664	0.4671	0.4678	0.4686	0.4693	0.4699	0.4706
1.9	0.4713	0.4719	0.4726	0.4732	0.4738	0.4744	0.4750	0.4756	0.4761	0.4767
2.0	0.4772	0.4778	0.4783	0.4788	0.4793	0.4798	0.4803	0.4808	0.4812	0.4817
2.1	0.4821	0.4826	0.4830	0.4834	0.4838	0.4842	0.4846	0.4850	0.4854	0.4857
2.2	0.4861	0.4864	0.4868	0.4871	0.4875	0.4878	0.4881	0.4884	0.4887	0.4890
2.3	0.4893	0.4896	0.4898	0.4901	0.4904	0.4906	0.4909	0.4911	0.4913	0.4916
2.4	0.4918	0.4920	0.4922	0.4925	0.4927	0.4929	0.4931	0.4932	0.4934	0.4936
2.5	0.4938	0.4940	0.4941	0.4943	0.4945	0.4946	0.4948	0.4949	0.4951	0.4952
2.6	0.4953	0.4955	0.4956	0.4957	0.4959	0.4960	0.4961	0.4962	0.4963	0.4964
2.7	0.4965	0.4966	0.4967	0.4968	0.4969	0.4970	0.4971	0.4972	0.4973	0.4974
2.8	0.4974	0.4975	0.4976	0.4977	0.4977	0.4978	0.4979	0.4980	0.4980	0.4981
2.9	0.4981	0.4982	0.4982	0.4983	0.4984	0.4984	0.4985	0.4985	0.4986	0.4986
3.0	0.4986	0.4987	0.4987	0.4988	0.4988	0.4989	0.4989	0.4989	0.4990	0.4990
3.1	0.4990	0.4991	0.4991	0.4991	0.4992	0.4992	0.4992	0.4992	0.4993	0.4993
3.2	0.4993	0.4993	0.4994	0.4994	0.4994	0.4994	0.4994	0.4995	0.4995	0.4995
3.3	0.4995	0.4995	0.4996	0.4996	0.4996	0.4996	0.4996	0.4996	0.4996	0.4996
3.4	0.4997	0.4997	0.4997	0.4997	0.4997	0.4997	0.4997	0.4997	0.4998	0.4998
3.5	0.4998	0.4998	0.4998	0.4998	0.4998	0.4998	0.4998	0.4998	0.4998	0.4998
3.6	0.4998	0.4998	0.4999	0.4999	0.4999	0.4999	0.4999	0.4999	0.4999	0.4999
3.7	0.4999	0.4999	0.4999	0.4999	0.4999	0.4999	0.4999	0.4999	0.4999	0.4999
3.8	0.4999	0.4999	0.4999	0.4999	0.4999	0.4999	0.4999	0.4999	0.4999	0.5000

For all $x \geq 3.89$, Area $= 0.5000$

Index

191